COMPUTER NETWORKS SERIES

IP PACKET FORWARDING RESEARCH PROGRESS

COMPUTER NETWORKS SERIES

File-Sharing Applications Engineering
Luca Caviglione
20009. ISBN 978-1-60741-594-7

Topology Construction for Bootstrapping Peer-to-Peer Systems over Ad-Hoc Networks
Wei Ding
2009. ISBN: 978-1-60692-919-3

IP Packet Forwarding Research Progress
Pi-Chung Wang
2009.ISBN: 978-1-60741-016-4

IP PACKET FORWARDING RESEARCH PROGRESS

PI-CHUNG WANG

Nova Science Publishers, Inc.
New York

© 2009 by Nova Science Publishers, Inc.

All rights reserved. No part of this book may be reproduced, stored in a retrieval system or transmitted in any form or by any means: electronic, electrostatic, magnetic, tape, mechanical photocopying, recording or otherwise without the written permission of the Publisher.

For permission to use material from this book please contact us:
Telephone 631-231-7269; Fax 631-231-8175
Web Site: http://www.novapublishers.com

NOTICE TO THE READER

The Publisher has taken reasonable care in the preparation of this book, but makes no expressed or implied warranty of any kind and assumes no responsibility for any errors or omissions. No liability is assumed for incidental or consequential damages in connection with or arising out of information contained in this book. The Publisher shall not be liable for any special, consequential, or exemplary damages resulting, in whole or in part, from the readers' use of, or reliance upon, this material.

Independent verification should be sought for any data, advice or recommendations contained in this book. In addition, no responsibility is assumed by the publisher for any injury and/or damage to persons or property arising from any methods, products, instructions, ideas or otherwise contained in this publication.

This publication is designed to provide accurate and authoritative information with regard to the subject matter cover herein. It is sold with the clear understanding that the Publisher is not engaged in rendering legal or any other professional services. If legal, medical or any other expert assistance is required, the services of a competent person should be sought. FROM A DECLARATION OF PARTICIPANTS JOINTLY ADOPTED BY A COMMITTEE OF THE AMERICAN BAR ASSOCIATION AND A COMMITTEE OF PUBLISHERS.

Library of Congress Cataloging-in-Publication Data

Wang, Pi-Chung.
　IP packet forwarding research progress / author Pi-Chung Wang.
　　　p. cm.
　Includes bibliographical references and index.
　ISBN 978-1-60741-016-4 (softcover)
　1. Packet switching (Data transmission) I. Title.
　TK5105.3.W36 2009
　004.6'6--dc22
　　　　　　　　　　　　　2009003382

Published by Nova Science Publishers, Inc. ✤ *New York*

Contents

Preface		**vii**
1	**Introduction**	**1**
	1.1. IP Address Lookup	3
2	**IP Address Lookup: A Brief Overview**	**7**
3	**IP Address Lookup with Hardware Pipelining**	**9**
	3.1. Related Works	9
	3.2. NHA Construction Algorithm	11
	3.3. NHA Compression Algorithm	14
	3.4. Hardware Implementation	16
	3.5. Performance Analysis	17
	3.6. Summary	18
4	**Performance Enhancement of IP Forwarding Using Routing Interval**	**21**
	4.1. Routing Interval	21
	4.1.1. Qualitative Analysis and Enhancement with Existing Routing Schemes	26
	4.2. IP Address Lookup by Using Routing Interval	28
	4.2.1. Routing Interval Transformation	28
	4.2.2. Modification of Binary Search for Speedup	32
	4.2.3. Further Improvement with Memory Bus Alignment	34
	4.2.4. Complexity	34
	4.3. Performance Evaluation	34

	4.4. Summary	38

5 High Speed IP Lookup Using Level Smart-Compression — 39
- 5.1. Trie-Based Algorithms ... 39
  - 5.1.1. Patricia Trie ... 39
  - 5.1.2. Level Compression Trie ... 40
  - 5.1.3. Lulea Compressed Trie ... 41
- 5.2. Level Smart-Compression Tries ... 42
  - 5.2.1. Level Smart-Compression Tries ... 43
  - 5.2.2. Route Update ... 48
- 5.3. Core-Leaf Decoupling ... 49
- 5.4. Performance Analysis ... 50
- 5.5. Summary ... 52

6 Fast Packet Classification by Using Tuple Reduction — 53
- 6.1. Introduction ... 53
- 6.2. Previous Works ... 55
- 6.3. Tuple Reduction Algorithm ... 58
  - 6.3.1. Filter Expansion ... 58
  - 6.3.2. Two-Dimensional Tuple Reduction to Minimize Storage ... 60
  - 6.3.3. Semi-Optimization Algorithm ... 61
  - 6.3.4. Tuple Reduction in Area of Sparse Filter ... 62
  - 6.3.5. Further Optimizing of Lookup Speed ... 63
- 6.4. Look-Ahead Caching Mechanism ... 64
  - 6.4.1. Realization of Parallel Hardware ... 66
- 6.5. Performance Evaluation ... 68
- 6.6. Summary ... 74

7 Conclusion — 77

Index — 83

Preface

Within the Internet, there exists a packet-switch network which functions to forward packets from source to destination in order to enable end-to-end connections. The key attribute to enable such a procedure consists of several crucial components, including routing protocol(s), transmission links and routers. While a plethora of literature on routing protocols has been presented in the last decade, the transmission technology is constantly evolving. As a result, provision of tens gigabit fiber links is commonly available now. Yet, the research on high-speed routers is limited insofar and, therefore, increases in its importance. To meet the demands of new multimedia applications, multi-tera routers have been designed. A multi-tera router should have enough internal bandwidth to switch packets between its interfaces at multi-tera rates and enough packet processing power to forward multiple millions of packets per second (MPPS). Switching in the router has been well studied. However, the remaining major bottleneck for a high performance router design is to speed up the multi-memory-access IP packet forwarding engine.

The rate of the packet forwarding engines can be burdened by several factors. The major obstacle lies in the space limitation of IPv4 which leads to the occurrence of classless interdomain routing (CIDR). The address prefix, specifying next-hop for a set of addresses, transforms from fixed length to variable length. Therefore, the packet forwarding engine must be able to derive the best matching prefix (BMP) among the variable-length prefixes. Unlike exact matching, the procedure of BMP requires more pre-computation to enable fast search. The second obstacle would be the incoming IPv6. Although IPv6 can overcome the problem of exhausting IPv4 address space, its huge space also increases the complexity of BMP search. Other factors affecting the speed of packet forwarding engines include the limited storage of high-speed memory, power consumption and the issues of table updates. Since there is no single solution fit under every circumstance, how to select a suitable solution for different applications is thereby important.

To solve these difficulties, numerous algorithms are proposed in the last few years. These algorithms address these performance issues via different approaches. For example, some of them try to compress data structures into high-speed SRAMs, while others utilize the architecture of modern processors. There are some others who resort to hardware implementation. Generally, these algorithms can be categorized into software based and hardware based according to the method of implementation. In this chapter, an introduction of four algorithms will be provided in terms of their major contributions, performance, merits and weaknesses.

Chapter 1

Introduction

The Internet is a global computer network that began as the ARPANET, an U.S. Department of Defense project to create a nation-wide computer network that would continue to function even if a large portion of it were destroyed in the war or natural disaster. However, the Internet was changed in 1992, when commercial entities started offering Internet access to the public. In addition, the Internet is used to provide enterprise communications. As personal computers became more affordable, people also began to connect to the Internet through **Internet Service Providers (ISPs)**. Due to the advance of the World Wide Web and the promise of future e-commerce, we have seen that the traffic in the Internet doubles every 3 months [1]. The number of hosts on the Internet has been growing exponentially, as shown in Figure 1.1. In addition, the introduction of the Gigabit Ethernet for LANs and the audio/video traffic generated by the multimedia applications increase the demand for higher bandwidth on the Internet. For example, the required bandwidth for high definition TV (HDTV) quality video is 20 to 40 Mbps or 2 to 6 with MPEG compression. When the network has to support a large number of such connections, the aggregate bandwidth requirement is high.

Internet consists of different subnetworks which are interconnected via gateways, routers, switches and various transmission facilities. **Internet Protocol (IP)** provides the functionality for connecting end systems across multiple networks which may use different protocols. This message (called **packet**) transfer process is called **packet forwarding**. The packets traverse from source to destination through links and routers, it is just like the way a postal mail traverses from post office to post office through delivery channels. Each post office uses

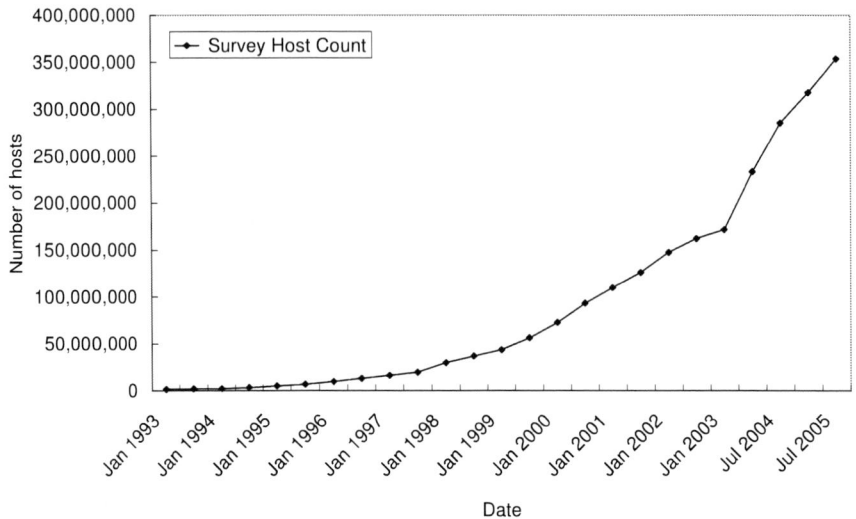

Figure 1.1. The Number of Hosts on the Internet.

the destination address on the envelope to decide where to forward this mail. In the same way, routers decide to forward a packet based on a destination address (**DA**) that is placed in the packet header, this process is called IP address lookup (or IP lookup, for short). Thus routers can forward packets through the Internet all the way to their destinations.

In addition to the basic packet forwarding based on destination address, the new requirements are emerging. To support security, quality of server (**QoS**) or specific business policies, user traffic may be further classified according to a maximum of eight fields: source/destination IP address (32 bits), source/destination transport-layer port numbers (16 bits for TCP and UDP), type-of-service (**TOS**) field (8 bits), protocol field (8 bits) and transport-layer flags (8 bits) with a total of 120 bits. This new forwarding type is called Layer 4 Switching because routing decisions can be based on headers available at Layer 4 or higher in the **Open Systems Interconnection** (**OSI**) architecture. Layer 4 switching offers increased flexibility: it gives a router the capability to block traffic from a dangerous external site, to reserve bandwidth for traffic between two specific sites or to give preferential treatment to one kind of traffic over other kinds. Traditional routers do not provide service differentiation because they treat all traffic in the same way. The process of mapping packets to different service classes is referred to as *policy-based routing*.

Speeding up the packet forwarding in the Internet backbone requires high-speed transmission links and high performance routers. The transmission technology keeps evolving and provision of gigabit fiber links is commonly available. For example, the high speed fiber links up to OC-192 (10 Gbps) are commercial available. Consequently, the key to increase the capacity of the Internet lies in fast routers [2]. A multi-gigabit router must have enough internal bandwidth to switch packets between its interfaces at multi-gigabit rates and enough packet processing power to forward multiple millions of packets per second (**MPPS**) [3]. Switching in the router has been studied extensively and solutions for fast packet processing are very popular. As a result, the remaining major obstacle for the high performance router design is the slow, multi-memory-access *IP/policy lookup* process.

In the past few years, new protocols such as tag switching and IP switching have been proposed to avoid the need for doing address lookups. This is because earlier algorithms for address lookups were too slow. The IETF has also instituted the Multiprotocol Label Switching (MPLS) [4] charter to study the use of label switching. Label switching is a generalization of tag and IP switching. MPLS proposes using labels to avoid address lookups and policy-based routing, it tends to achieve better traffic engineering. However, the MPLS efforts primarily focus on achieving better traffic engineering, as opposed to avoiding address lookups or policy-based routing because the average utilization of Internet backbone networks is estimated at 15% [5]. However, these solutions cannot completely avoid the need for IP lookups, thus it still requires fast address lookup algorithms for both protocols.

1.1. IP Address Lookup

The size of a IPv4 packet header is 20 bytes without extensions. It contains the source and destination addresses. An IP address consists of a network identifier and a host identifier. Routing is solely based on the network identifier. Originally, the network identifier had a predetermined length, indicated by a prefix in certain address bits which specified the address class: 0 indicated class A with 8-bit network identifiers, 10 indicated class B with 16 bits, and 110 indicated class C with 24-bit identifier. The class-based structure are proved too inflexible and wasteful of the address space. As a result, it is outmoded by the introduction of classless interdomain routing (CIDR) [6] in 1993.

With CIDR, IP routes have been identified by a ⟨routing prefix, prefix length⟩ pair, where the prefix length varies from 1 to 32 bits. Arbitrary aggregation of networks are allowed to reduce routing table entries, as well as increase the utilization of address space. Suppose all the subnets in a big network have identical routing information except for a single, small subnet that has different information, rather than having multiple routing entries for each subnet in the large network, just two entries are needed: one for the big network and a more specific one for the small subnet. The entry-length distribution of a sample CIDR routing table from the network access point (**NAP**) Mae-East is shown in Figure 1.2. Although the class-based address structure has been abolished, it is still visible in routing table entries since the numbers of prefixes with length 16 and 24 bits (formerly class B and C, respectively) are more than other lengths, especially the 24-bit prefixes. Although the CIDR results in better usage of the available IP address space as well as decreases the number of routing table entries, it also increases the processing power requirement for doing IP lookup.

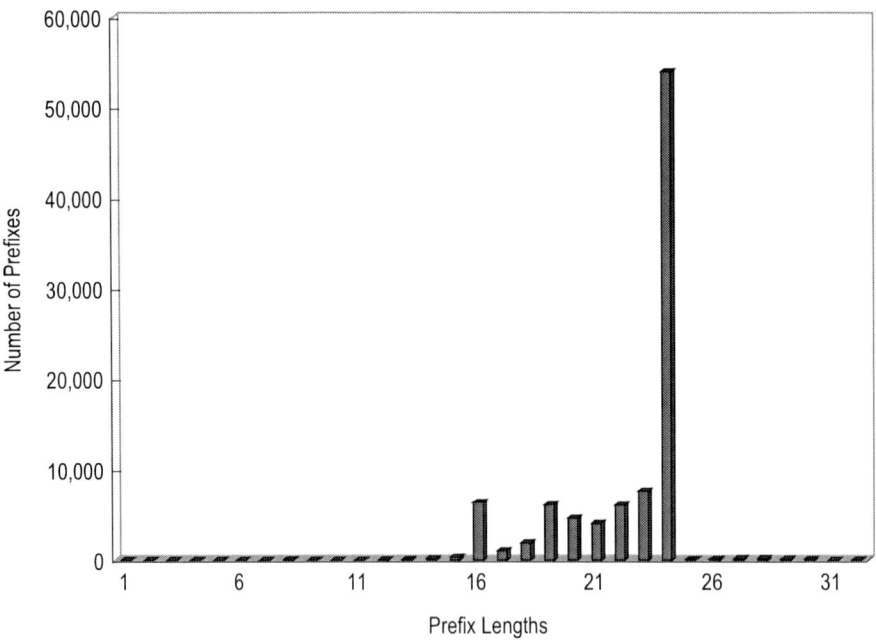

Figure 1.2. The Prefix Length Distribution.

An IP address can be split into network and host identifiers from any point. The network identifier is the prefix that is stored in the routing table. For instance, the address 140.113.215.213 gives the network identifier 140.113.208 with 20-bit prefix and 140.7 with 12-bit prefix. An address that would match both prefixes in a router should consequently be routed according to the information kept for the longer one, which is 20-bit prefix in this case. We call this **Best Matching Prefix (BMP)** problem.

To solve the BMP problem, traditional exact-matching lookup algorithms, such as hash and CAM, cannot be applied trivially due to two facts: 1) the prefixes length is variable and 2) there are multiple matching prefixes to the same DA. Thus it is time-consuming to search for the BMP, especially in a backbone router with a large number of table entries.

Most routers have a default route that is given by a prefix of length zero and therefore matches all addresses. The default route is used consequently if no other prefix matches. The backbone routers in the Internet are required to recognize all network identifiers and cannot resort to the default route. Thus their routing table is larger than other routers. Although the address structure for IPv6 is not public available, there are suggestions to keep the variable-length network identifiers. As result, the longer address length of IPv6 will only compound the problems of routers.

The result of an IP-address lookup in a router is the port number and next-hop address that should be used for the packet. The next-hop address is used to find the physical-link address for the next downstream router when the interconnection is via a shared-medium network (e.g., Ethernet address). The next-hop is not needed for point-to-point links, and the corresponding routing-table entry would only contain the output port number. Even when next-hop addresses are needed, they are much fewer addresses than the entries in the routing table. One can use an 8-bit pointer pointing to an array that lists the next-hop addresses.

Chapter 2

IP Address Lookup: A Brief Overview

There has been a remarkable interest in the organization of routing tables during the past few years. The proposals can be separated into hardware and software based solutions.

- Hardware-based Solutions

    For years, designers of fast routers use caching to claim high speed IP lookups. This is problematic for several reasons, potentially diluting the cache with hundreds of addresses that map to the same prefix. Second, a typical backbone router of the future may have hundreds of thousands of prefixes and be expected to forward packets at tera-bit rates. Although studies have shown that caching in the backbone routers can result in hit ratios up to more than 90 percent [7,8], the simulations of cache behavior were done on large, fully associative caches which are commonly implemented using CAMs. CAMs, as already mentioned, are usually expensive. It is not clear how set associative caches will perform and whether caching will be able to keep up with the growth of the Internet. So caching does help, but does not avoid the need for fast BMP lookups, especially in view of current network growth.

    Besides cache schemes, Degermark *et al.* [9] proposed a trie-like data structure. A major concern in their work is the size of the trie to ensure that it fits in on-chip cache memory. The main idea of their work is to quantify the prefix lengths to levels of 16, 24 and 32 bits and expand each prefix in the table to the next higher level. Using this structure, it is able to compact a large routing table with 40,000 entries into a table with 150-160 Kbytes size. If implemented in hardware, the minimum and maximum number of memory accesses for a lookup are two and nine, respectively. Gupta *et al.* presented fast routing-lookup schemes based on a huge DRAM [10]. The scheme accomplishes a maximum of two memory accesses for a lookup in a forwarding table of 33 megabytes. By adding an intermediate-length table, the forwarding table can be reduced to 9 megabytes; however, the maximum number of memory accesses for a lookup is increased to three. When implemented in a hardware pipeline, it can achieve one route lookup every memory access. This furnishes about 20 MPPS. Huang *et al.* [11] further improve it by fitting the forwarding table into SRAM.

- Software-based Solutions

    Regarding software solutions, algorithms based on tree, hash or binary search have been proposed. Srinivasan *et al.* [12] present a data structure based on binary tree with multiway branching. By using a standard trie representation with arrays of children pointers, insertions and deletions of prefixes are supported. However, to minimize the size of the tree, dynamic programming is needed. In [13], Karlsson *et al.* solve the BMP problem by *LC tries* and linear search. Waldvogel *et al.* proposed a lookup scheme based on a binary search mechanism [14]. This scheme scales very well as the size of address and routing tables grows. The scheme requires a worst-case time complexity of log_2(address bits) hash lookups. Thus, five hash lookups are needed for IPv4, and seven for IPv6 (128-bit). This software-based binary search work is further improved by employing a cache structure as well as using multiway and multicolumn search techniques [15]. For a database of N prefixes with address length W, the native binary search scheme needs $O(W \times logN)$ searches. This improved scheme takes only $O(W + logN)$ searches.

In the following, we present three algorithms for IP address lookup. One is suitable for hardware implementation, one for software implementation and the other for both.

Chapter 3

IP Address Lookup with Hardware Pipelining

First, we present a BMP scheme for hardware implementation in [16]. Based on the implementation, the forwarding table would be reduced to enough small to fit into SRAM with very low cost. For example, a large routing table with 53,000 entries can be compressed to a forwarding table of 540 Kbytes. Most of the address lookups can be accomplished in a single memory access. In the worst case, the number of memory accesses for a lookup is two. When implemented in pipelined hardware, the scheme can achieve one address lookup per memory access. With state-of-the-art SRAM technology, this mechanism can achieve more than 100 million lookups per second.

3.1. Related Works

Several works have been proposed with novel data structures to reduce the complexity of longest-prefix matching lookups [9, 12–15]. These data structures and their accompanying algorithms are designed primarily for software implementation, they are usually unable to complete a lookup in one memory-access-time. The most straightforward way to implement a lookup scheme is to have a forwarding table for each IP address, as depicted in Figure 3.1. However, the size of the forwarding table (next-hop array; NHA) is too large (4 GBytes) to be practical.

To reduce the size of the forwarding table, an indirect lookup mechanism,

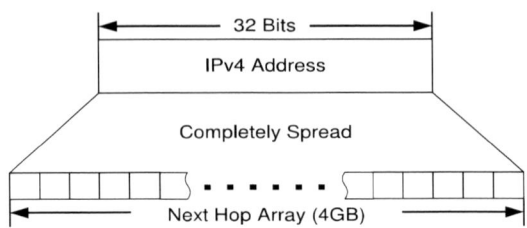

Figure 3.1. Direct-lookup mechanism.

as shown in Figure 3.2 [10], can be employed. Each IP address is split into two parts: 1) segment (16-bit) and 2) offset (16-bit). The segment table has 64K entries which record either the next hop of the route or pointers pointing to the associated **Next-hop Array** (**NHAs**). Each NHA consists of 64K entries, each entry records the next hop for the destination IP address. This scheme uses a maximum of two memory accesses for a lookup in a 33-Mbyte forwarding table. By adding an intermediate-length table, the forwarding table can be reduced to 9 Mbytes; however, the maximum number of memory accesses for a lookup will be increased to three. When implemented in hardware pipeline, the scheme can accomplish one address lookup every memory access and achieve up to 20 million lookups per second.

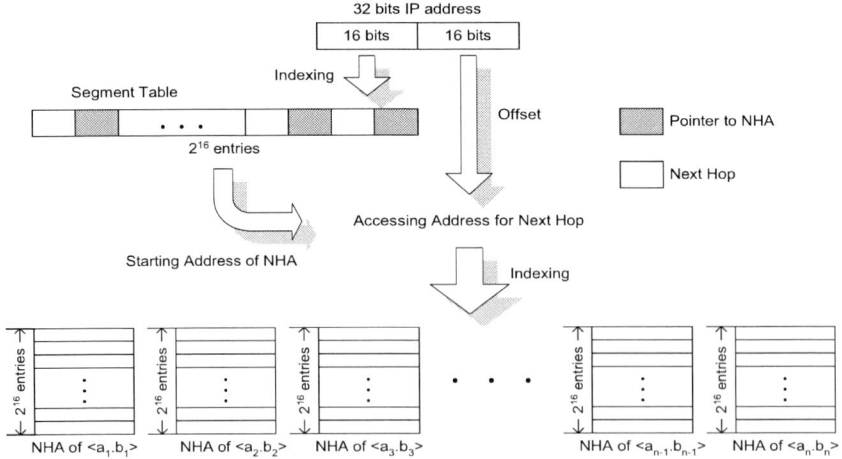

Figure 3.2. Indirect-lookup mechanism.

IP Address Lookup with Hardware Pipelining

In [11], Huang *et al.* further reduced the size of the associated NHA by considering the distribution of the prefixes belonging to the same segment, as demonstrated in Figure 3.3. With 45,000 routing prefixes, it results in about 750K entries in the NHA. With 1 byte per entry, the total required memory size (including segment table) is more than 1 MBytes, and the required number of memory accesses is two. By employing the concept of compression, the required memory size can be further reduced, but the number of memory access would be increased to three. The time complexity for building the so-called **Code Word Array (CWA)** and the **Compressed NHA (CNHA)** is $O(n\log n)$ [11], where n denotes the number of prefixes in a segment. However, due to the mechanism of compression, it is needed to rebuild the CNHA for updating the forwarding table. Since the routing updates may occur every few seconds, the performance might degrade severely due to the memory bandwidth contention.

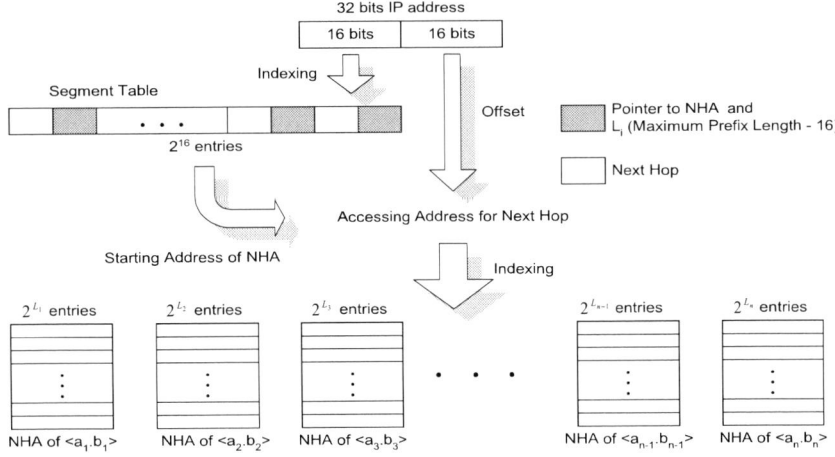

Figure 3.3. Indirect-lookup mechanism with variable offset length.

3.2. NHA Construction Algorithm

By observing the sample routing table in Figure 3.4, one can find that all routing prefixes are belonging to the same segment $\langle 63.192 \rangle$. In Huang's algorithm, the total required entries in an NHA would be $2^{(20-16)} = 24$. After a further analysis of these prefixes, one can find that the first 17 bits $\langle 00111111110000000 \rangle$ of these prefixes are the same. Since the longest prefix length is 20, this means that only the 3-bits variation is needed to build an NHA with $2^{(20-17)} = 2^3$-entries for these 6 prefixes, as shown in Figure 3.4. The longer the common part is, the shorter the NHA entries will be. The only problem is how to represent the extra prefix information. To deal with this issue, few prefix bits and its length are added to the entry of the segment table. This will increase the size of the segment table as a trade off. Those entries with no routing prefix information, such as P_5, should be filled with a default route.

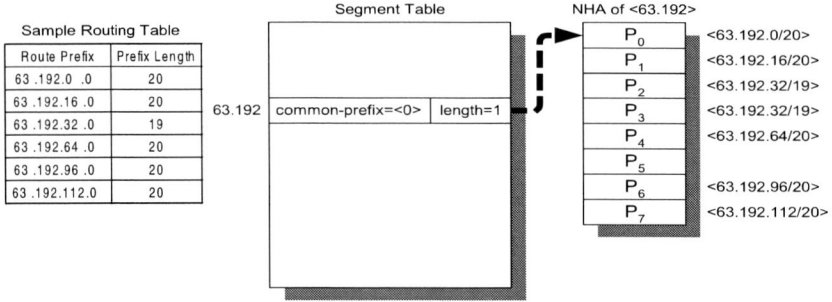

Figure 3.4. The resulting NHA from the sample routing prefixes.

The NHA construction algorithm for a segment is given bellow. Let l_i and h_i be the length and output port identifier of a routing prefix p_i, respectively. The *cprefix* represents the common part of routing prefixes beyond the first 16 bits in a segment, and *clength* is the valid length of *cprefix*. *Mlength* equals to the longest prefix length in the segment minus 16, thus, $clength \leq Mlength$. Let $p_i(x,y)$ represent the bit pattern of p_i from the x_{th} bit to the y_{th} bit. The entries from $V(p_i(clength + 16, l_i)) \times 2^{Mlength-(l_i)-16}$ to $V(p_i(clength + 16, l_i)) \times 2^{Mlength-(l_i)-16} + (2^{Mlength-(l_i-16)} - 1)$, where $V(p_i(a,b))$ represents the value of bit pattern $p_i(a,b)$. Besides, most routers have a default route with zero prefix length which matchel addresses. The default route is used consequently if no other prefix matches. Thus the table is initially assigned the default route for possible reference to these entries.

NHA-Construction Algorithm
Input: The set of routing prefixes of a segment.
Output: The corresponding NHA of this segment.

Step 1. Let l_i and h_i be the length and output port of a routing prefix p_i, respectively.

Step 2. Let $P = \{p_0, p_1, ..., p_{n-1}\}$ be the set of sorted prefixes of an input segment. For any pair of prefixes p_i and p_j in the set, $i < j$ if and only if $l_i < l_j$.

Step 3. Let $cprefix = p_0(17, i_0)$, $clength = l_0 - 16$ and $Mlength = l_{n-1} - 16$.

Step 4. For $i = 0$ to $n - 1$ do
  $cprefix$=common bits between $cprefix$ and $p_i(17, l_i)$.
  $clength$=valid length of $cprefix$.

Step 5. Construct the NHA with $2^{Mlength-clength}$ entries and set an initial value to the default route.

Step 6. For $i = 0$ to $n - 1$ do
  Calculate the range of updated entries and set h_i.

Step 7. Stop.

Let us use an example to show how this algorithm works. Consider the set of sorted prefixes in Figure 3.5. In Step 4, all prefixes are examined once for deciding the *cprefix* and *clength* value, and it results in a single bit "0" for *cprefix* and thus *clength* is 1. Since *Mlength* is equal to 8, the constructed NHA is with $2^{8-1} = 128$ entries. After constructing the NHA, the table is assigned with default route as the initial value because the routing prefixes cannot cover every entry in NHA. Then the first prefix $\langle 24.48.8/22/10 \rangle$ will be fetched. The $8_{th}(= 2 \times 2^{8-6})$ to $11_{th}(= 2 \times 2^{8-6} + 2^{8-6} - 1)$ entries are the associated entries for $\langle 24.48.8/22/10 \rangle$ and will be overwritten with the output port value 10. The process is repeated for other prefixes. When processing the prefix $\langle 24.48.9/24/7 \rangle$, one can find that the value of 9_{th} entry is 10, which is defined by prefix $\langle 24.48.8/22/10 \rangle$, but the prefix $\langle 24.48.9/24/7 \rangle$ is a longer one. To satisfy the longest prefix matching, the 9_{th} entry will be overwritten with 7 again. Since the prefixes have been sorted by their lengths, this process can be done trivially.

Obviously, the computation cost is low. If a new routing prefix is received, it will recalculate *Mlength*, *cprefix* and *clength*, which is same as Step 4, and rebuilds the NHA if its size is changed. By applying the algorithm, the number

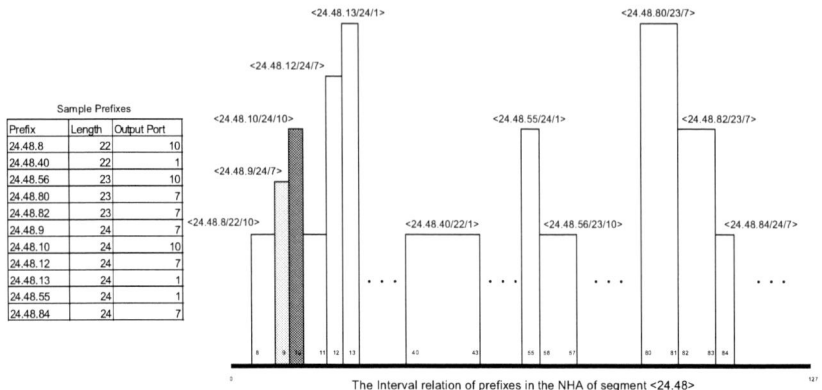

Figure 3.5. The NHA construction example.

of entries of the generated NHA can be 25% less than Huang's algorithm. The detailed organization of the segment table and the hardware architecture will be addressed below as well as the effect of the *cprefix*.

3.3. NHA Compression Algorithm

In the NHA construction algorithm, the size of the generated forwarding table would be around 1.2 Mbytes with one byte per entry which includes 3-bit *cprefix*. Moreover, the NHA can be further compressed based on the distribution of output port count within the NHA. The distribution, which is generated from traces of several main NAPs, is shown in Figure 3.6. It is observed that approximately 54% segments consist of less than two output ports and 96% segments use less than four output ports. Thus fewer bits are enough to encode the output port identifier in an NHA. Two extra fields, *cbits* and *base*, are appended to the entry of the segment table. At first, a bit-vector is used to record all possible output port identifiers, which are then encoded using the smallest port identifier as the *base*. Each associated NHA entry will be set to the value of its output port identifier minus the *base*. Also, the required bits are set to *cbits* for each entry. Thus the *base* can be read before indexing the associated NHA, and the physical output port index can be calculated by adding the base index to the NHA indexing result. This process can be completed within two memory accesses plus the calculation time which can be ignored. the detailed compressed-NHA construction algorithm is listed below.

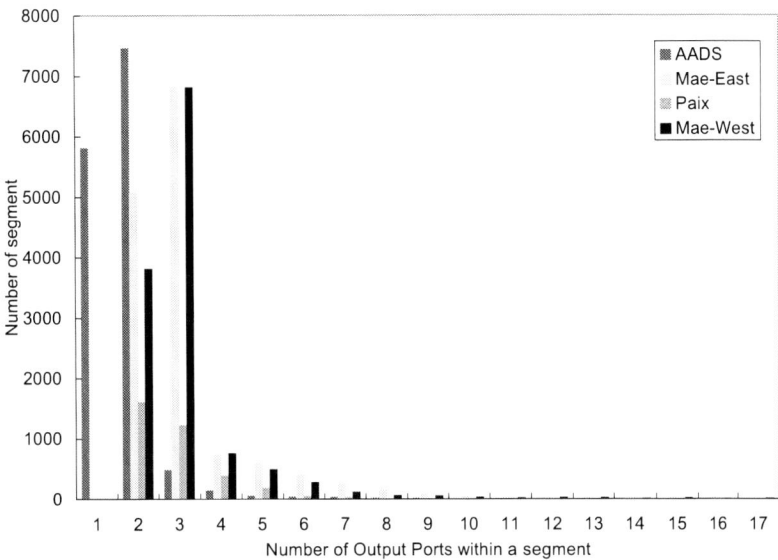

Figure 3.6. The distribution of output port count within a segment.

Compressed-NHA Construction Algorithm
Input: The set of routing prefixes of a segment.
Output: The corresponding compressed NHA of this segment.

Step 1. For each prefix, calculate *cprefix*, *Mlength* and *clength*. Also, records the output port value in the output port vector.

Step 2. Construct the index table from the output port vector and calculate the *cbits* and *base* value according to the output port count.

Step 3. Construct the NHA with size $2^{Mlength-clength} \times cbits$.

Step 4. For each routing prefix
Calculate the index value of an output port by subtracting the base value.
Calculate the range of updated entries in NHA and fill it with index value.

Step 5. Stop.

Although the table compression ratio in the scheme is not so notable, it is able to accomplish an IP lookup with two memory accesses. Also the scheme is implementation feasible with pipelining hardware. Notice that in Huang's algorithm, the memory locations accessed in the second and the third lookup are the same because it appends the CNHA to the tail of the CWA. Therefore, it may cause the structural hazard for implementation in pipelining hardware. Such potential structural hazard can be avoided in the scheme without requiring any specific hardware such as dual port memory.

3.4. Hardware Implementation

A feasible high-level hardware architecture of the lookup scheme is shown in Figure 3.7. When a DA is fetched, its first 16 bits are used as an index to the segment table. The value of pointer/next_hop field in the corresponding entry records the next hop if it is smaller than 256. The value will be forwarded to the selector directly. Otherwise, it records the starting address of the associated NHA. Then the bit pattern *DA(clength, Mlength)* is used to compare with the *cprefix* stored in this entry. If two values are identical, then *V(DA(clength, Mlength))* plus the value of the pointer is used to access the NHA. The physical output port can be derived by adding the *base* with the value fetched from NHA. Otherwise, the default output port will be forwarded to the selector.

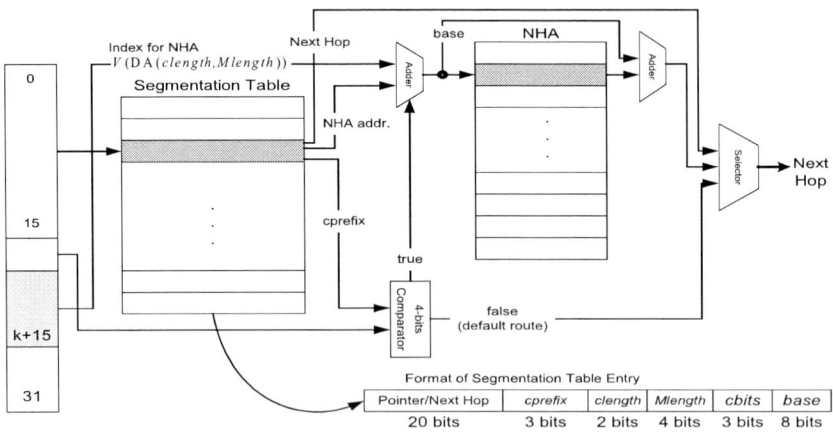

Figure 3.7. Hardware architecture of the lookup scheme.

The entry of the segment table consists of 6 fields: pointer/next_hop, *cprefix*, *clength*, *Mlength*, *base* and *cbits*. The length of pointer/next_hop is 20 bits which can map to 1 mega memory addresses, this is more than five times the space required for storing the whole NHAs generated from the real world routing tables. Since the maximum prefix length minus the length of segment (16) is smaller than 16, the length of *Mlength* is 4 bits. Trivially, the base is 8 bits and the maximum number of encoded bits is 8. Thus the bit count of *cbits* is 3. If the length of *cprefix* is 3, then the *clength* is 2 bits. As a consequence, the length of each entry is 40 bits, and the size of the segment table is $2^{16} \times 40$ bits, i.e., 320 Kbytes.

3.5. Performance Analysis

The IP address lookup scheme aims at two targets, one is to reduce the memory space needed for the forwarding tables, and the other is to speed up the IP lookup process. Note that the current backbone routers have a routing table with about 53,000 entries. We use the logs of publicly available routing tables as the basis for comparison. These tables are offered by the IPMA project [17], they provide a daily snap shot of the routing tables used by some major NAPs.

To further realize the effect of the extra *cprefix*, we use the trace available in the router of Mae-East NAP to present the relation between the total number of entries in NHAs and the length of *cprefix*, as shown in Figure 3.8. By coupling the effect of *cprefix*, the number of entries can be 25% less than that required in Huang's algorithm. Although as the length of *cprefix* increases, the number of NHA entries is reduced, it has to increase the entry length of the segment table, thus results in a larger segment table. From the observation of Figure 3.8, the suitable *cprefix* length should be no more than 3 by considering both segment table size and the number of NHA entries.

In Table 3.1, we log five traces to build the forwarding table for illustrating the effect of the scheme. Although the size of the forwarding table might become larger as a trade-off for throughput improvement, we can find that the required memory size does not increase too much, or even decrease in two traces (PacBell and Paix). This is because that the use of *cprefix* reduces the number of NHA entries effectively. While in the traces of two backbone NAPs (Mae-East and Mae-West), the memory increment is notable, this is due to the diverse routing prefixes in the backbone, in which the effect of *cprefix* is degraded.

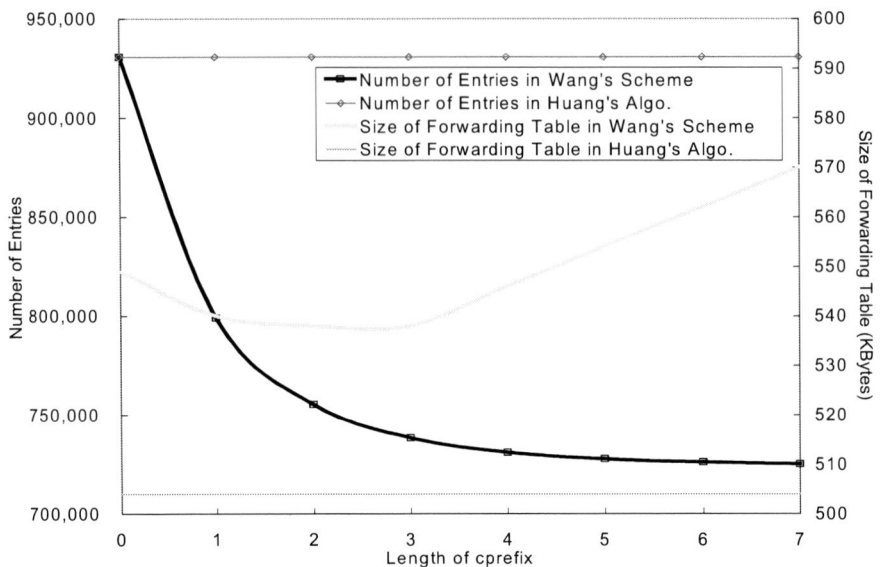

Figure 3.8. Effect of the Length of *cprefix*.

3.6. Summary

In this section, we introduce a fast IP lookup scheme that requires less memory space, and it can be implemented with high-speed SRAM. By employing an efficient compression scheme, the size of the forwarding table can be reduced significantly. For example, a large routing table with 53,000 routing entries can be compressed to a forwarding table of 540 Kbytes with the cost of less than US$30. Most of the address lookups can be done in one memory access,

Table 3.1. The comparison of memory requirements.

Site	Routing Prefixes	Wang's Scheme (Kbytes)	Huang's Algo (Kbytes)
AADS	23,426	396	403
Mae-East	53,226	538	504
PacBell	29,942	258	412
Paix	11,498	328	351
Mae-West	33,332	529	454

with the worst case, two. Furthermore, the scheme can avoid the structural hazard for hardware pipelining. When implemented in hardware with pipeline technique, it can accomplish one address lookup in a single memory access and achieve approximately 100 million address lookups per second with current 10ns SRAM.

Chapter 4

Performance Enhancement of IP Forwarding Using Routing Interval

Nowadays, the commonly used table lookup scheme for IP routing is based on the so-called classless interdomain routing (CIDR). With CIDR, routers must find out the BMP for IP packets forwarding, this complicates the IP lookup. Currently, this process is mainly done in software and several schemes have been proposed for hardware implementation. Since the IP lookup performance is a major design issue for the new generation routers, the properties of the routing table are investigated and a new approach for IP lookups is presented in [18, 19]. This approach is not based on BMP and thus significantly reduces the complexity. By applying the approach, the computation cost of existing schemes can be significantly reduced. Moreover, the performance of the binary search on prefixes is improved to 30 MPPS and 5,000 route updates per second under the same experiment setup with an even larger routing table. With the state-of-the-art 700 MHz CPU, it can achieve more than 70 MPPS in average.

4.1. Routing Interval

According to the concept of CIDR, each IP address might be covered by multiple routing prefixes. However, only the longest one indicates the next-hop. From the geographical characteristic of the destination location, a new routing concept based on the following observations is presented. Before presenting our approach, it is necessary to define the segment for IP address. Here each IP address can be divided into two parts: segment (16 bits) and offset (16 bits).

The density of routing prefixes for segments is related to its hop-count: From the network topology, all networks (including enterprise, campus, ISP) are physically interconnected through trunks. Since there are multiple trunks for neighbor cloud to connect with each other, the number of routing prefixes is large and with disorderly information. On the other hand, the routing prefixes are simple and much fewer for the remote cloud, i.e., farther geographical region.

The number of routing prefixes is sparse in IP network: Although the number of Internet hosts increases exponentially, the distribution of routing prefix is still very sparse. For example, there are 2^32 IP addresses but only about 56,000 routing prefixes in current backbone router (Mae-East 8/12/2000). There is less than one routing prefix per segment on average. We use the routing tables offered by the IPMA project [17], which provides a daily snap shot of the routing table used by some major NAPs, as basis for our experiment. The number of prefixes whose length is longer than 16 bits in each segment is shown in Figure 4.1. From the experimental results, we observed that most of routing prefixes identify the next-hop only for few segments, this phenomenon is called as routing locality. It also reflects the property that most prefixes are related to neighbor clouds.

In a router, the number of possible next-hops for a segment is always much less than the total number of ports: The number of distinct next-hops in a routing table is limited by the number of other routers or hosts that can be reached in one hop, it is clear that these numbers would be small even for large backbone routers. For the enterprise or campus routers, the extensive use of default routes for outside destinations further reduce the routing table and the number of next-hops. It results in that few routing prefixes are used to define the routes for most segments. Hence the number of next-hops for most segments is small and hence less than the number of related routing prefixes, as shown in Figure 4.2.

From the above discussion, a new routing concept "Routing Interval", which fully utilizes the aforementioned properties, is introduced. By sorting the routing prefixes based on their lengths, a next-hop array contains entries, which correspond to all intervals. In addition, each entry also records the related next-hop. For example, there are three routing

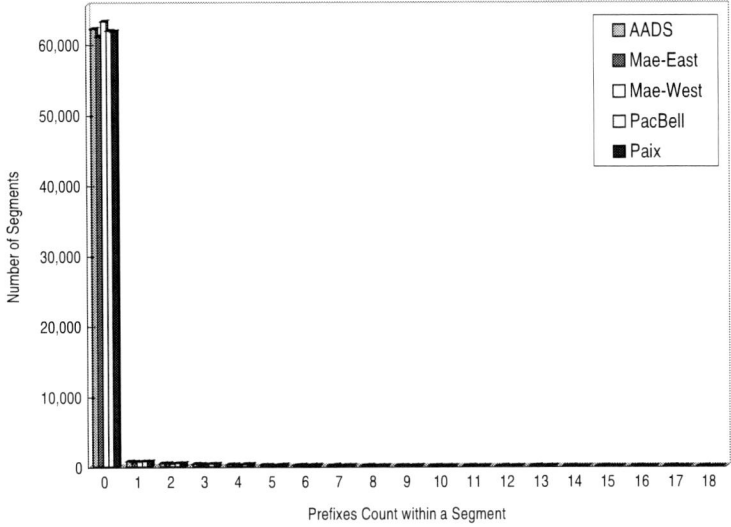

Figure 4.1. Prefix Count Distribution.

prefixes $140.113.0.0/255.255.0.0/NH_1$, $140.113.3.0/255.255.255.0/NH_2$ and $140.113.0.0/255.255.255.0/NH_3$ with length 16, 24, 24, respectively. First, the routing prefix $140.113.0.0/255.255.0.0/NH_1$ creates an IP address interval $140.113.0.0 \sim 140.113.255.255/NH_1$. When the next routing prefix $140.113.3.0/255.255.255.0/NH_2$ is processed, the interval is further partitioned into three regions, $140.113.0.0 \sim 140.113.2.255/NH_1$, $140.113.3.0 \sim 140.113.3.255/NH_2$ and $140.113.4.0 \sim 140.113.255.255/NH_1$, respectively. According to this simple rule, three routing prefixes results in five intervals based on the next-hop value and record each with $\langle begin\ address, end\ address, next-hop \rangle$. This tuple is named as "Routing Interval", as shown in Figure 4.3.

Most of existing algorithms use different approaches to handle the BMP problem and result in either complex data structure or extra storage, or both. The most significant benefit to transform the routing prefix into routing interval is that the complexity of BMP problem can be removed. For example, each node in Patricia trie consists of three pointers, one for prefix entry and two for child nodes due to possible longer match. Although various implementations of the Patricia trie have been proposed and significantly improve the performance, it can be further enhanced via applying the routing interval. The inefficient data

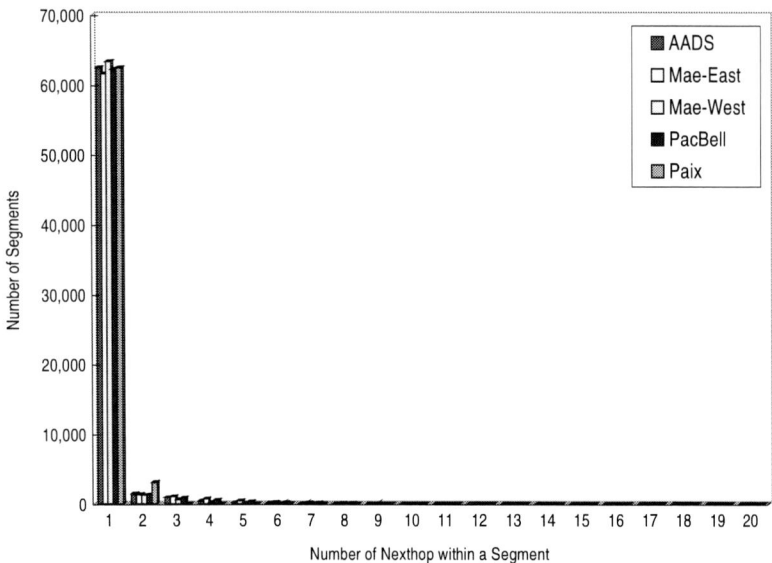

Figure 4.2. The Distribution of Next-hop Count within a Segment.

structure can be simplified, with which each node may contain either a prefix pointer or two child pointers, but not both. With low transformation cost, the BMP problem is reduced into a much simpler search problem. The algorithm of transformation is addressed below. Notice that the idea is completely different from the previous work [20], which concerns how to reduce the number of routing prefixes, while the concept of routing interval focuses on how to remove the BMP problem.

Theoretically, the maximum number of generated routing intervals is $2N - 1$, where N is the number of routing prefixes (including default route). However, the experiments show different results. The result generated from the large routing table (Mae-East) is shown in Figure 4.4. The number of routing intervals is less than that of routing prefixes in this case. There may be three reasons: (I) A region which indicates identical next-hop value has to be represented by several routing prefixes due to the limitation that the occupied region of routing prefixes must be with the order of 2. (II) Some routing prefixes belong to the same routing interval, thus they can be merged into a single interval. For example, there are two routing prefixes, $140.113.0.0/16/NH_1$, $140.113.3.0/24/NH_1$, with difference prefix length and identical next-hop value. The routing prefix

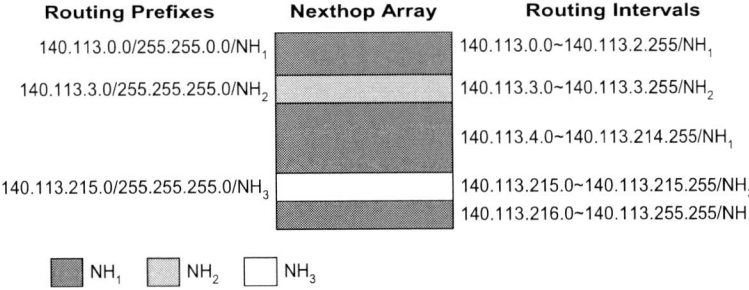

Figure 4.3. Routing Interval Example.

$140.113.3.0/24/NH_1$ is redundant and can be removed. (III) There are zero-range intervals which can be eliminated. This phenomenon can be understood after the discussion in Section 4.2.1.. The results of other NAPs are shown in Table 4.1, where the value below is the number of routing prefixes. In most of the numerical results, the generated numbers of intervals are fewer than that of the routing prefixes.

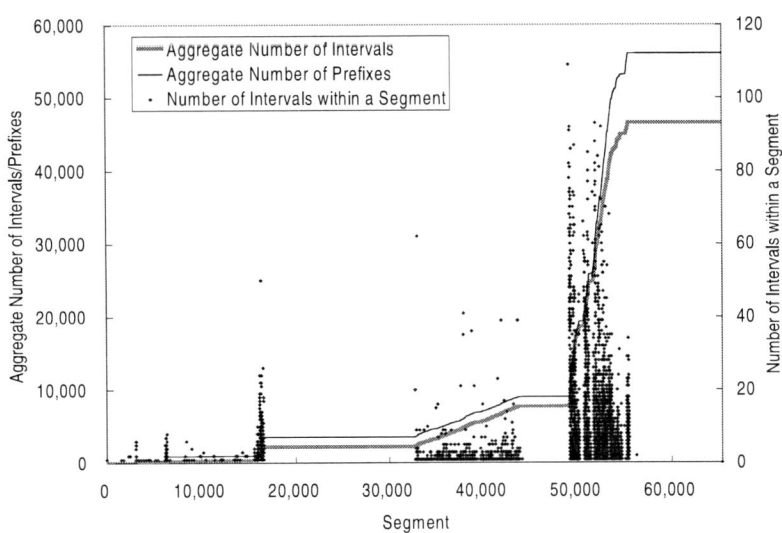

Figure 4.4. Number of Generated Routing Intervals for Mae-East.

For most of the segments, the routing information, including the next-hops

Table 4.1. Comparison of Routing Prefixes/Routing Intervals in Different Dates.

Date	AADS	Mae-East	Mae-West	PacBell	Paix
9/9/1999	9,287 35,769	44,233 47,784	28,853 28,998	22,263 26,278	9,341 9,270
12/22/1999	12,075 17,460	44,904 52,372	30,387 30,688	21,141 25,406	8,711 9,237
5/15/2000	24,451 23,515	42,800 53,226	30,492 33,514	23,619 29,959	8,944 11,587
8/12/2000	23,411 26,109	46,557 56,039	30,501 31,060	25,807 33,680	14,201 13,878

and routing prefixes, is very simple. Further, the distribution is related to the network topology. The routing information is more complex for the nearby region (in topology). Therefore, it is possible to improve the IP lookup scheme based on these characteristics by converting the hierarchical routing prefixes into flat routing intervals. With the concept of routing interval, complexity of IP lookup can be greatly reduced. Next, how the existing schemes can take advantage of this concept is shown in the following.

4.1.1. Qualitative Analysis and Enhancement with Existing Routing Schemes

The scheme proposed by Waldvogel *et al.* [14] performs the IP lookup based on the lengths of the prefixes. The prefixes with the same length are stored in a hash-table. The binary search is used to find out the longest prefix, and one hash-table lookup is performed at each step of this search. However, to find out the BMP with binary search is not trivial. For each prefix with length L, an extra marker will be added to the hash table with length 1 to $L-1$ to indicate that there exists a longer match, so that it will lookup the longer-prefix hash table. Furthermore, the best match prefix for the marker should also be calculated to avoid misleading. While the routing prefixes is replaced by the routing intervals, it does not need to change the basic mechanism, but the pre-computation of best match for each marker can be removed. Since no misleading will happen with the usage of routing interval, there is no backtracking problem. Thus, the pre-computation cost can be saved while simplifying the data structure of forwarding table.

To migrate the routing prefixes to routing intervals here, the routing intervals must satisfy that the region of each interval can be represented as the power of two. For example, an interval whose region is six will be cut into two intervals with region value four and two respectively. A sample result converted from Figure 4.4 is shown in Figure 4.5. The number of the processed intervals would be a trade off between simplified construction and searching algorithm.

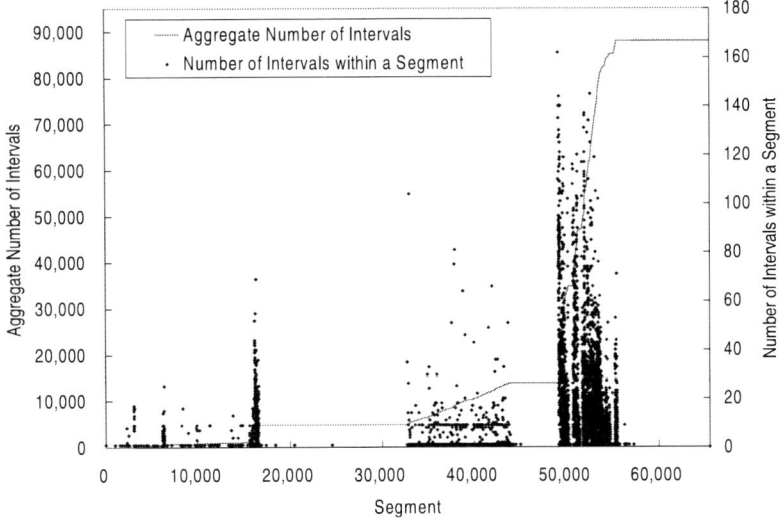

Figure 4.5. Number of Processed Routing Intervals for Mae-East.

Another scheme proposed by Srinivasan *et al.* [12] consists of two parts. In the first part, it reduces the number of distinct prefix lengths. With dynamic programming, the optimal expansion level can be calculated. Consequently, it constructs the trie based on the results. Leaf pushing and cache line alignment are used to improve the performance. While applying routing intervals, the operation of leaf push can be eliminated because the longer match will not occur. Each element in the node indicates a pointer to either the next node or the prefix table. In fact, leaf pushing is conceptually similar to our idea, but ours is more generalized. In [12], the authors address how to improve the performance of binary search on hash table. By applying the routing interval concept, the enhancement discussed above can be applied, too.

In [13], Nilsson *et al.* use the skill called level compression tries (*LC tries*) to reduce the height of trie. The basic idea is: since there is a fixed cost for comparing strings whose lengths are equal or less than the machine word, it is therefore more efficient to compare more bits at a time to reduce the number of comparisons and memory accesses. However, to improve the efficiency of data structure, they didn't handle the BMP problem in the trie. Each internal node keeps the number of nodes and the address of the leftmost one. The search only stops at the leaf node, which contains the pointer to the prefix table. Once the pointed prefix doesn't match, it will jump to the shorter prefix of current one indicated in this entry. Therefore, it deals the BMP problem with linear search, this might result in poor performance in some cases.

Applying routing interval can eliminate the aforementioned problem. If the trie is constructed with routing intervals, each interval will correspond to a leaf node in the trie. Once the search reaches the leaf node, the best matching will be found. Since no linear search is needed, the prefix table can be removed.

4.2. IP Address Lookup by Using Routing Interval

We have introduced three enhancements to previous schemes and show the advantages of using the routing interval. Basically, it can simplify the data structure and reduce the pre-computation cost that leads to low table construction time. It might also reduce the required memory storage and the number of accesses in some cases. The transformation from routing prefix to routing interval is essential to design an effective routing scheme. Consequently, an efficient transformation method is presented as well as an IP lookup scheme, which further takes advantage of routing intervals. The algorithm can handle 5,000 route updates per second, which includes the interval transformation and forwarding table construction.

4.2.1. Routing Interval Transformation

According to the routing interval concept, each routing prefix implies a routing interval that can be identified by just appending (32-prefix length) 0's and (32-prefix length) 1's as starting and ending address, respectively. For example, routing prefix $P_i = \langle 140.113.0.0/16/NH_i \rangle$ is with a routing interval $\langle S_i = 140.113.0.0, E_i = 140.113.255.255, NH_i \rangle$. However, it is an imprecise routing interval since it overlaps with other routing interval in the boundary that causes the ambiguous address lookup. Consequently, the major purpose of routing interval transformation is used to partition the routing intervals precisely.

The insertion of a routing prefix will generate at most three routing intervals which are $\langle unknown, S_i - 1, unknown \rangle$, $\langle S_i, unknown, NH_i \rangle$ and $\langle E_i + 1, unknown, unknown \rangle$. The transformation algorithm is used to determine these unknown values. Basically, the transformation from routing prefix to interval is irreversible because the hierarchical information carried in routing prefix is removed. Each route update will cause the re-transformation of the routing interval. To make it easier for implementation, two link-lists, L_1 and L_2, are adopted. The former is used to store the generated routing interval formats, and it is initialized to empty. The latter is used to store the possible routing interval $\langle E_i + 1, unknown, unknown \rangle$ and it is implemented as a stack, which performs "push", "pop" and "modification" operations.

With the sorted routing prefixes, let S_i, E_i and NH_i be the starting address, ending address and the next-hop of the routing interval P_i, respectively, where $S_1 \leq S_2 \leq \ldots \leq S_N$ (N is the number of routing prefixes, $1 \leq i \leq N$). At the beginning, we append the interval $\langle 0, unknown, default\ route \rangle$ to L_1 and push the interval $\langle 255.255.255.255, 255.255.255.255, default\ route \rangle$ into L_2. To simplify the description, we use the subscript top and $rear$ to represent the starting address, ending address and next-hop of the top and rear routing interval in L_1 and L_2, respectively. For each routing prefix with imprecise routing interval $\langle S_i, E_i, NH_i \rangle$, we first check whether S_{top} (i.e., the starting address of the top interval in L_2) is smaller than S_i, if so, $E_{rear} = S_{top} - 1$. Then, we pop the top element, say, $\langle S_{top}, E_{top}, NH_{top} \rangle$, of L_2 and append it to the rear of L_1. After that, we update NH_{rear} with NH_{top}. We repeat these steps until there is no smaller interval in L_2, i.e., $S_{top} \leq S_i$, and modify E_{rear} to $S_i - 1$. It means that there exist an interval $\langle S_{rear}, S_i - 1, NH_{rear} \rangle$. Moreover, we must check whether S_{rear} is equal to S_i or not. If yes, then we replace NH_{rear} with NH_i. Otherwise, we append $\langle S_i, unknown, NH_i \rangle$ to L_1. Since it belongs to an interval which is not defined clearly, we must further check whether S_{top} is equal to $E_i + 1$ or not. If $S_{top} = E_i + 1$, then we replace NH_{top} with NH_i. Otherwise, we push an interval, $\langle E_i + 1, unknown, NH_i \rangle$, into L_2. Finally, we pop all intervals stored in L_2 and append them to L_1 sequentially. The detailed algorithm is listed as below.

Routing Interval Transformation Algorithm

Input: N routing prefixes.
Output: The set of ordered routing intervals.
Let S_i and E_i be the starting and ending address, NH_i be the next-hop of i_{th} routing prefix, respectively. (The subscript *top* and *rear* represent the starting address, ending address and the next-hop of the *top* and *rear* interval in L_1 and L_2, respectively.)
Let $P = P_1, P_2, \ldots, P_{N-1}$ be the set of sorted prefixes of an input segment.
For any pair of prefixes P_i and P_j in the set, $i < j$ if and only if $S_i < S_j$.
Append interval $\langle 0, unknown, default\ route \rangle$ into L_1 and push interval $\langle 1111\ldots 1, 1111\ldots 1, default\ route \rangle$ into L_2.
For $i = 1$ to N **do**

1. For the imprecise interval $\langle S_i, E_i, NH_i \rangle$ of i_{th} routing prefix. Check if S_i larger than S_{top}.

2. If yes, modify E_{rear} with $S_{top} - 1$. Pop the top interval from L_2 into L_1. Then modify NH_{rear} with NH_{top}. Repeat step 1&2 until the result of comparison is false.

3. Consequently, modify E_{rear} as $(S_i - 1)$.

4. Check if $S_{rear} = S_i$,
   4.a If yes, overwrite NH_{rear} with NH_i.
   4.b Otherwise, append $\langle S_i, unknown, NH_i \rangle$ into L_1.

5. Check if the $S_{top} = E_i + 1$,
   5.a If yes, overwrite NH_{top} with NH_i.
   5.b Otherwise, push $\langle E_i + 1, unknown, NH_i \rangle$ into L_2.

Push all intervals stored in L_2 and append into L_1.
For all generated intervals **do**
     Merge two successive intervals with identical next-hop.
Stop.

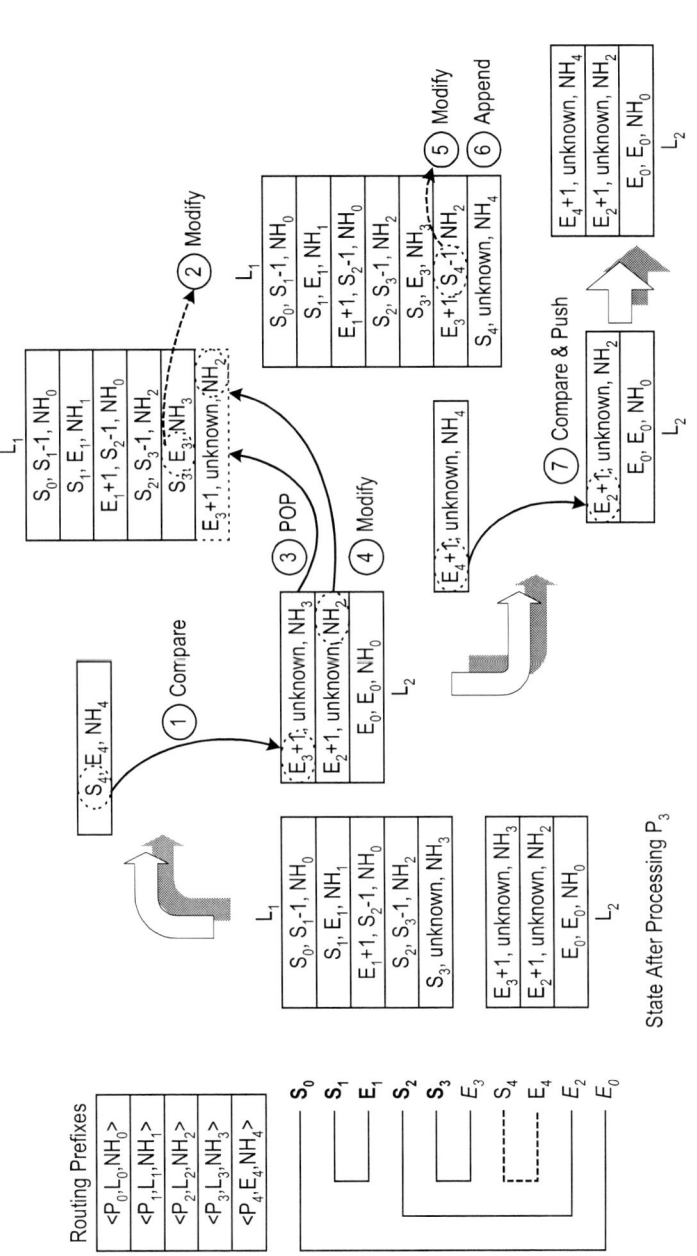

Figure 4.6. An Example of Routing Interval Transformation.

In Figure 4.6, we use an example to illustrate the process of the routing interval transformation. After processing the routing prefix P_3, the top element of L_2 and rear elements of L_1 are $\langle E_3, unknown, NH_3 \rangle$ and $\langle S_3, unknown, NH_3 \rangle$, respectively. For the routing prefix P_4, it is firstly compared with S_{top}. Since S_{top} is smaller than S_4, E_{rear} is modified to S_i. It pops the top element, $\langle E_3+1, unknown, NH_3 \rangle$, of L_2 and append the popped element to the rear of L_1. Then we modify NH_3 with NH_2. Since there is no smaller interval in L_2, we replace E_{rear} with $S_4 - 1$. It further checks whether S_{rear} is equal to S_4 or not. Because they are different, it must append the interval $\langle S_4, unknown, NH_4 \rangle$ to L_1. In addition, S_{top} is unequal to E_4+1, thus it pushes the interval $\langle E_4+1, unknown, NH_4 \rangle$ into L_2. After processing all prefixes, the ordered intervals are listed in the array. The time complexity of the transformation algorithm without the prefixes sorting is $O(M) = O(2N-1) = O(N)$ where M is the interval count and N is the prefix count.

4.2.2. Modification of Binary Search for Speedup

In [15], Lampson *et al.* use two copies of the routing prefixes, one copy is padded with zeros and the other with ones, for IP lookup. With pre-computation, it can perform the prefix matching with binary search in a sorted array containing these extended prefixes. The required memory of forwarding table is about 1 Mbytes, which cannot fit into the L2 cache of most modern CPUs. Recall that a CPU READ to a byte will prefetch an entire cache line into the L2 cache, multiway search with cache line alignment are used to improve the performance. In spite of its scalability to IPv6, the route update will cause the reconstruction of the forwarding table, which costs about 350 ms in the worst case.

The performance can be greatly improved by adopting the routing interval. The mapping of previous scheme from routing prefix to routing interval is shown in Figure 4.7. From the right part of Figure 4.7, one can see that a native binary search can support IP lookup without any further pre-computation. Moreover, the number of entries (M) is always less than that of previous scheme ($2N$). However, the problem of route updates remains, there is no existing update technique that is faster than just building a table from scratch. Another problem resulted from the unbalanced interval distribution, as shown in Figure 4.1, might degrade the efficiency of binary search.

To alleviate both problems, an IP address is split into two parts: the segment (16-bit) and the offset (16-bit), and add a segment table which extract the pre-

Performance Enhancement of IP Forwarding Using Routing Interval 33

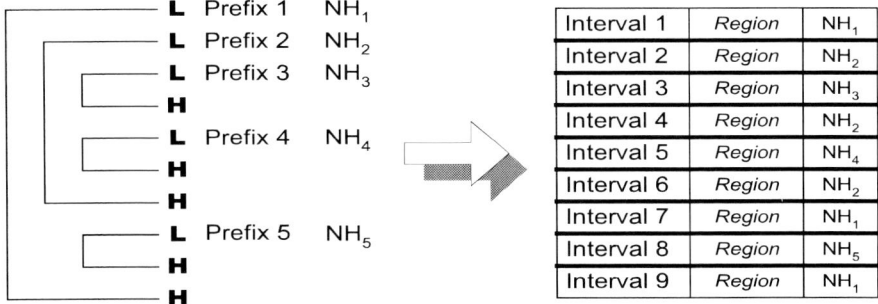

Figure 4.7. Mapping from Previous Scheme into the Routing Interval Scheme.

fixes with length less than 16, as used in [15]. Each entry of the segment table consists of two fields: address of the first interval and the number of intervals, as shown in Figure 4.8. If the interval count is zero, the value in another field indicates the next-hop. Otherwise, it will perform a binary search to the intervals whose first 16 bits are the same. With 2 bytes for each field, the size of segment table will be 256 Kbytes.

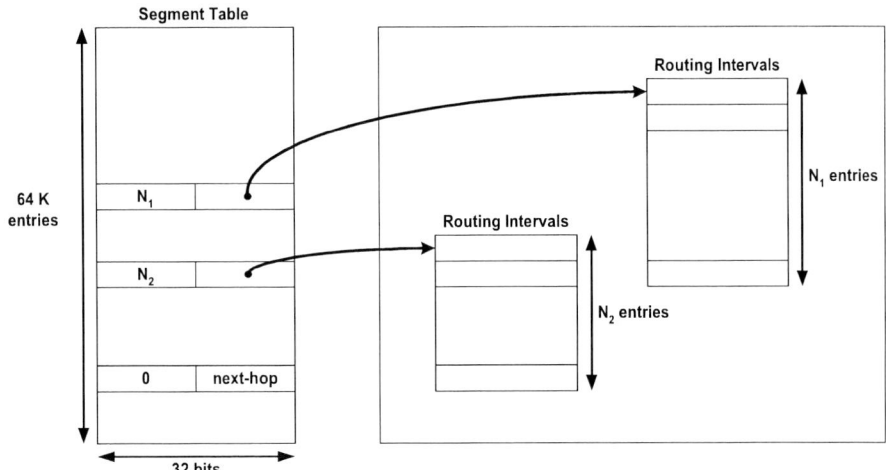

Figure 4.8. Binary Search With 16-bits Segment Table.

4.2.3. Further Improvement with Memory Bus Alignment

To achieve better lookup performance, two requirements should be met. At first, the size of forwarding table should be small enough to fit into the L2 cache. By constructing the forwarding table, the total size including segment table is less than 480 Kbytes for Mae-East routing table. Consequently, the data structure for binary search should comply with the memory bus. Unlike the cache line alignment described in the previous scheme, our improvement focus on the characteristic of memory bus between L1 and L2 cache. Most modern CPUs have large bus width, such as 64-bits, for access efficiency. Therefore, two intervals (5-bytes for each) is merged into one entry with 8 bytes, as shown in Figure 4.9. The binary search will test two routing intervals within one L2 cache access. The size of forwarding table can be reduced and the lookup speed can be fastened.

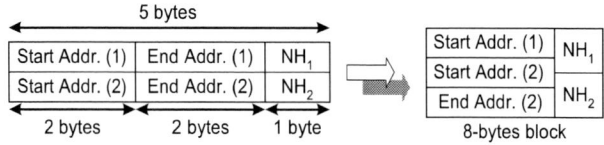

Figure 4.9. Memory Bus Alignment for Access Efficiency.

4.2.4. Complexity

In Table 4.2, we show the complexity required for different software-based schemes. Note that these complexity measures do not indicate anything about the physical speed or actual memory usage. In the worst case, the number of routing intervals is twice of the number of routing prefixes. We will show the realistic performance metrics in the next section.

4.3. Performance Evaluation

To simplify the comparison with multiway search, we use a comparable platform as used in previous work [15]. We choose a 300-MHz Pentium II running Windows NT that has a 512-Kbytes L2 cache. We also use another 700-MHz K7, state-of-the-art high performance CPU, to show what performance level we

Table 4.2. Comparison with the Existing Works.

Algorithms	Build	Search	Memory
Patricia Trie	O(NW)	O(W)	O(NW)
Binary Search on Hash Tables	O($Nlog_2W$)	O(log_2W)	O($Nlog_2W$)
Multibit Trie	O(hW^2)	O(h)	O(hN)
LC Tries	O(hN)	O(h)	O(hN)
Multiway Search	O(N)	O($log_M(2N)$)	O($2N$)
Routing Interval Scheme	O(N)	O($log_2(2N)$)	O($2N$)

N: the number of the prefixes, W: the length of the address,
h: the height of the trie, M: the number of branches

can achieve. Five routing tables from the IPMA project are used for the experiment. We will show the performance of the scheme with respect to three metrics: worst/average search time, storage and construct/update time.

The detailed performance metrics are shown in Table 4.3. The routing interval scheme works very well even with a large routing table. The update time represents the cost of table reconstruction for the segment with most intervals. Both construction and update time are accounted as the cost of interval transformation. The scheme can support at least 5,000 route updates per second which is much faster than 100 update/s as required in some BGP implementation [15]. The reason for the shortened build/update time is that our table construction is very simple. With link-list implementation, no large memory copy is needed in the interval transformation. Moreover, the construction of memory-bus-alignment array is much simpler than any existing schemes. The constructed forwarding table is small enough to fit into CPU L2 cache. Besides, the memory bus alignment helps improve the access efficiency as well as decrease the number of entries with a factor of 2. For the routing table of Mae-East, the worst-case forwarding rate is larger than 4 MPPS. If the weight of each IP address is the same, the average forwarding rate is about 30 MPPS. As used in practical environment, the performance would be better due to the locality of traffic flows.

Consequently, the routing interval scheme is compared to multiway search. The result generated from the routing table of Mae-East is shown in Table 4.4. Note that numbers measured on 200-MHz Pentium Pro have been proportionately scaled to 300 MHz, as used in [15]. The critical build/update problem

Table 4.3. Performance Evaluation with Five Routing Tables.

Performance Metrics	AADS	Mae-East	Mae-West	PacBell	Paix
Prefix Count	26,109	56,039	31,060	33,680	13,878
Interval Count	23,411	46,557	30,501	25,807	14,201
Memory Required (Kbytes)	374	459	393	372	327
Build Time (msec)	93	141	109	110	62
Update Time (μsec)	183.6	183.6	121.1	183.6	121.1
Average Lookup Time (nsec)	34	39	34	32	29
Worst Case Lookup Time (nsec)	244	244	244	244	244

is significantly improved. This is mainly because the pre-computation cost for solving the BMP problem is removed in the routing interval scheme. Moreover, the transformation from routing prefixes to intervals is performed cost effectively with link-list implementation. As a result, the build/update cost can be reduced significantly. Since no pre-computation information should be carried, the forwarding table is very simple and efficient. It can fit into the CPU L2 cache, thus results in better lookup performance. Therefore, even with slower L2 cache (150 MHz vs. 200 MHz), the worst-case lookup time is much shorter than that in multiway search.

Table 4.4. Comparison with Multiway Search.

Performance Metrics	Multiway Search	Routing Interval Scheme
Worst Case Lookup Time (nsec)	330	244
Worst Case Update Time (msec)	352	0.2
Build Time (sec)	5.8	0.14
Memory Required (Kbytes)	950	459

In Table 4.5, we further compare the routing interval scheme to other existing algorithms. Again, we proportionally scaled all results of previous works to 300 MHz CPU to ease the comparison. The worst-case lookup time of *LC tries* [13] is not addressed in the literature so we fill it with the average lookup time. Another important factor not shown in the table is the cost of route update which is not addressed in the papers on *LC tries*, binary search on hash tables, and Lulea compressed tries. However, all these schemes potentially require changing the complete data structure during a route update, they are likely to feature slow insertion/deletion times. Although the worst-case lookup time of multibit trie [12] is better than the scheme, we used a larger routing table for experiment (56,039 vs. 38,816). Also, their worst-case update time (2.5 msec) is much larger than ours.

Table 4.5. Comparison with Other Existing Works.

Previous Schemes	Worst Case Lookup Time (ns)	Memory Required (Kbytes)
Patricia Trie	1,650	3,262
Binary Search on Hash Tables	650	1,600
Lulea Scheme	409	160
Multiway Search	330	950
LC tries	(*)500	464
Multibit Trie	236	640
Routing Interval Scheme	244	459

(*)This is the average performance since the worst-case performance is not addressed.

From the results of performance evaluation, the routing interval scheme not only reduces the memory size significantly, outperforms the existing schemes in lookup speed, but also provides a much faster routing-table update. Furthermore, it just uses simple data structure and address lookup operations. Finally, we employ the state-of-the-art AMD 700 MHz CPU to investigate the possible highest performance of the scheme. As shown in Table 4.6, the average packet-forwarding rate is above 70 MPPS with 6 MPPS in the worst case. Thus, both build and update time benefit from the upgrade of CPU speed.

Table 4.6. Performance Evaluation with 700 MHz CPU.

Site	Required Memory (Kbytes)	Build Time (msec)	Update Time μsec	Average Lookup Time (nsec)	Worst Case Lookup Time (nsec)
Mae-East	459	70	117	22	153

4.4. Summary

In this section, we present a new concept named "routing interval" which can be used to resolve the BMP problem and significantly reduces the routing complexity. A routing interval transformation algorithm is introduced to support an efficient IP address lookup. Via experiment, both the table construction time and table size can be reduced significantly. The routing interval scheme not only reduces the memory size requirement significantly, outperforms the existing schemes in lookup speed, but also provide a fast routing-table update. The constructed forwarding table is small enough to fit into the CPU L2 cache. By using the fast table-reconstruction algorithm, it is able to provide more than 5,000 route updates per second. This feature is preferred to the backbone router for handling the frequent route updates. With memory bus alignment, the routing interval scheme can further achieve 70 MPPS with state-of-the-art 700 MHz CPU. These results demonstrate that the scheme is efficient and flexible for IP packets forwarding.

Chapter 5

High Speed IP Lookup Using Level Smart-Compression

In the following, we introduce a new trie-based data structure: *Level Smart-Compress tries* (*LS tries*). By using the new data structure, the size of the forwarding table can be compressed to less than 900 Kbytes for a large routing table with 100,000 routing entries. It can accomplish one IPv4 route lookup with a maximum of nine memory accesses and support incremental updates. The heuristics named core-leaf decoupling is presented to further improve the lookup performance and IPv6 scalability.

5.1. Trie-Based Algorithms

5.1.1. Patricia Trie

The trie is a general-purpose data structure for storing strings. The idea is very simple: each string is represented by a leaf in a tree structure, and the value of the string corresponds to the path from the root of the tree to the leaf. This simple structure is not very efficient. The number of nodes may be large and the average depth (the average length of a path from the root to a leaf) may be long. The traditional technique to overcome this problem is to use *path compression* when each internal node with only one child is removed. Of course, it has to somehow record those nodes which are missing. A simple technique is to store a number in each node, called the *skip* value, which indicates how many bits have been skipped on the path. A path-compressed binary trie is sometimes referred to as a Patricia trie, which is implemented in the IP lookup of NetBSD [21,22]. Consider the example shown in Figure 5.1. There are five bit streams in the left part. Based on the given routing table, the constructed Patricia trie is shown in the right part of Figure 5.1. Basically, it processes an address one bit at a time.

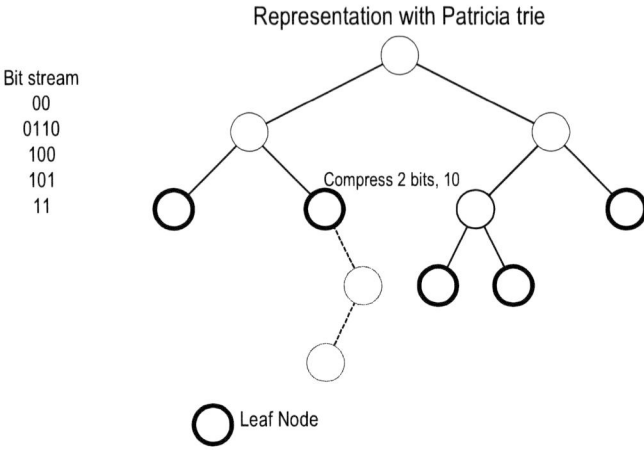

Figure 5.1. Representation of Patricia Trie.

5.1.2. Level Compression Trie

One might think of path compression as a way to compress the parts of the trie that are sparsely populated. Level compression [23] is a recently introduced technique for compressing parts of the trie that are densely populated. The idea is to replace the i highest complete levels of the binary trie with a single node of degree 2^i; this replacement is performed recursively on each subtrie. The level-compressed version of the trie in Figure 5.1 is shown in Figure 5.2. In Figure 5.2, one might find that there are eight leaves generated but only five route prefixes in the table. Accordingly, three nodes are over-generated in the trie, which is noted in Figure 5.2.

The performance of the LC-trie is much better than that of the Patricia trie since comparing strings of lengths equal to or less than the machine word size has basically a fixed cost, thus it is more efficient to compare more bits at a time in order to reduce the number of comparisons and memory accesses.

However, the implementation in [13] uses array representation by laying the

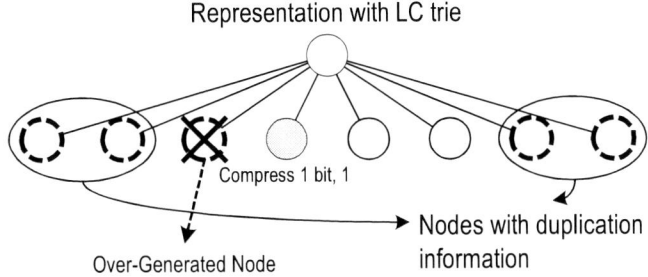

Figure 5.2. Representation of Level-compressed trie.

LC-trie nodes out in breadth first order (first the root, then all the trie nodes at the second level from left to right, then third level nodes etc.). Each node only carries the first address of its children. While performing address lookup, the procedure stops only when the leaf node is reached. Yet the fetched node may not be the BMP due to path compression. If the prefix in the node doesn't match the address, the shorter prefix of the node will be tracked until matched. Furthermore, the array layout and the requirement for full subtries make updates slow in the worst case. Further, it might cause almost every element in the array representation to be moved. Thus while LC-trie have reasonable average lookup speeds, they feature long insertion time and worst-case lookup performance.

The data structure presented in [12] is similar to LC-trie. It is also a binary trie structure, and it allows multiway branching. By using a standard trie representation with arrays of children pointers, insertions and deletions of prefixes can be supported. However, to minimize the size of the initial trie, complex dynamic programming is used.

5.1.3. Lulea Compressed Trie

The Lulea scheme compresses multibit trie nodes so that the entire data structure can now be placed in SRAM. In contrast with the Patricia/LC tries to compresses the trie structurally, Lulea scheme compresses the information in a node. The nodes are compressed by representing repeated elements in a node array only once and using a bitmap to specify the number of times an element gets repeated, as shown in Figure 5.3.

To speed up the counting of set bits, the algorithm accompanies each bitmap with a summary array that contains a cumulative count of the number of set bits

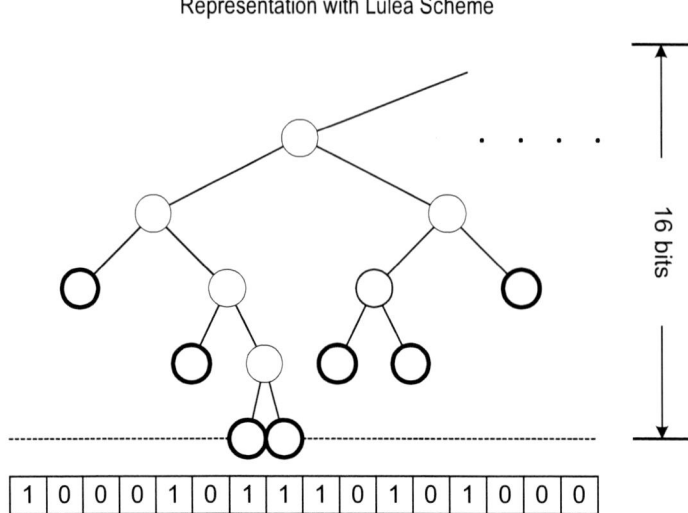

Figure 5.3. Representation of Lulea Compressed trie.

associated with fixed size chunks of the bitmap.

The Lulea scheme features compact storage but it has two disadvantages. First, counting bits requires at least one extra memory reference per trie node. Second, the implicit use of leaf pushing, makes the worst-case insertion time large. A prefix added to a root node entry can cause information to be pushed to thousands of leaf nodes.

5.2. Level Smart-Compression Tries

Figure 5.4 conceptually depicts the architecture of the IP router, which mainly consists of a network processor and a forwarding engine. The forwarding engine employs a forwarding table (FT) downloaded from the network processor to make the routing decision. The network processor executes the routing protocols, such as RIP and OSPF, and maintains a routing table (RT) and a dynamic forwarding table for fast updates. Once the route is updated, the forwarding table in the network processor will be renewed. Consequently, the forwarding engine will update its table based on the update information from the network processor without downloading a new forwarding table.

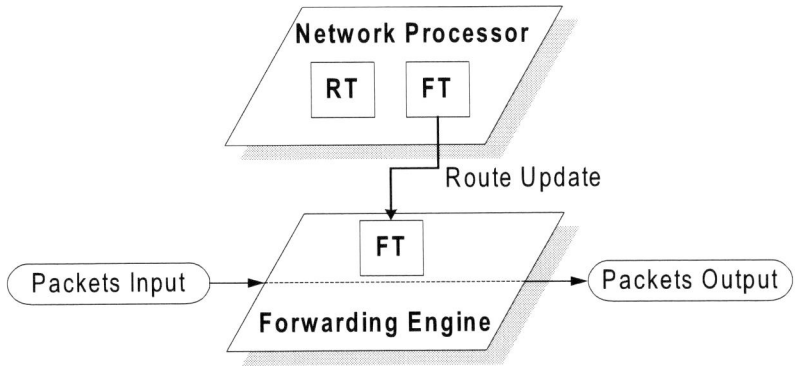

Figure 5.4. Architecture of the IP Router.

5.2.1. Level Smart-Compression Tries

If routing-table can be fit into the high speed SRAM, the performance can be promoted significantly. The data structure of level smart-compression tries (LS tries) is presented to achieve this goal. We use a simple binary tree as an example to explain the basic idea. As shown in Figure 5.5, the binary tree is constructed from the sample routing table. The bold circle indicates a node/leaf that identify a prefix. If the leaf node can be reached without traversing the internal node (thin one), the tree size can be further reduced, also the lookup performance can be increased. It is different to the ideas used in the Patricia tries and LC tries. In the former, only the one-child path is reduced, while in the later extra storage might be required to full expansion. With the scheme, only the necessary nodes are generated to greatly reduce the size of the forwarding table.

The data structure consists of the intermediate node and the leaf, as shown in Figure 5.6 and Figure 5.7, respectively. The value 2^N and 2^M indicate the number of branches which the intermediate node and the leaf will "administer", respectively. In the intermediate node, it consists of two parts: a head and a pointer. In the node head, the next-hop and length of the current BMP is recorded. It will be referenced only when the search is stopped as fetching a non-available value (NA) which indicates that there is no more specifically matched prefix. For example, P_4 will be used when search for "01001*" in Figure 5.5. It is called as a potential prefix since it is referred only when the destination address no longer match specific one. The deployment of this information can avoid the complex route update resulted from leaf-pushing, which

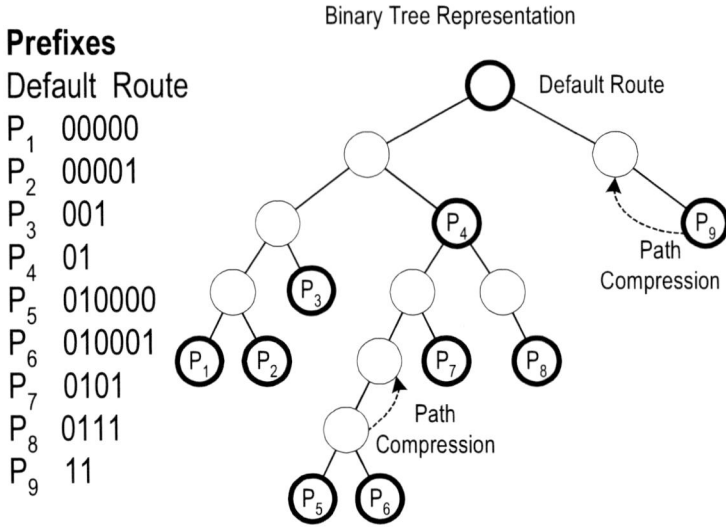

Figure 5.5. Representation of the Bit Stream With Binary Tree.

will be explained in the subsection 5.2.2.. The node pointer is mainly used to indicate the address of the children. The one-bit flag is used to identify whether a child is an intermediate node or a leaf. Since the position of the leaf is variable, an extra field with x bits is used to record its relative depth in the intermediate node. The node information and pointer are stored in two different arrays with one-to-one mapping and the later will be referred only when the non-available value is read in the node pointer or the leaf.

The data structure of the leaf is a next-hop array which consists of 2^M elements, as shown in Figure 5.7. It is generated by expanding M bits of the route prefixes.

Before constructing the forwarding table, a binary tree without path compression is generated from the original routing table. Then the binary tree is traversed with DFS. Firstly, the distance between the current node and the deepest node is checked whether it is less than or equal to M. If yes, a leaf is generated by expanding all routing prefixes included in the subtree. Otherwise, the left and right child nodes will be traversed to construct the forwarding table. In addition, it should decide whether to construct an intermediate node by dividing the current depth to N. If the result is zero, an intermediate node will be generated. The pseudo code of the construction algorithm is shown below. The

Node Head

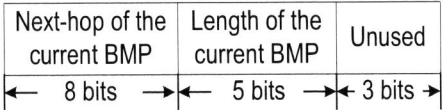

Node Pointer

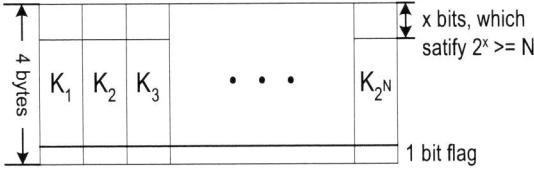

Figure 5.6. Intermediate Node of LS trie.

parameter "Parent" is used to identify the related intermediate node for the leaf.

LS trie Construction Algorithm
Input: The root of the binary trie which is constructed from the routing table.
Output: The forwarding table.
Constructor (Parent, Root, Depth, Prefix) {
  IF ((the deepest child depth in the subtree - Depth) is less or equal to M)
    Generate_Leaf (Parent, Root);
  ELSE {

Leaf

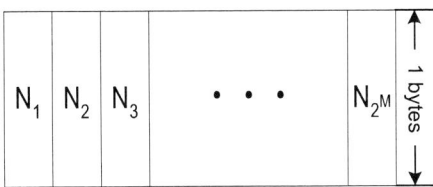

N_i: next-hop

Figure 5.7. Leaf of LS trie.

```
    If (Root− >Prefix not equal to NULL)
        Prefix=Root− >Prefix;
    If ((Depth mod N) is equal to 0)
        Parent=Generate_Intermediate_Node(Root,Prefix);
    Constructor(Parent, Root− >Left_Child, Depth+1);
    Constructor(Parent, Root− >Right_Child, Depth+1);
    }
}
```

By applying the algorithm to the binary trie in Figure 5.5, the constructed forwarding table is shown in Figure 5.8 ($N=3$, $M=3$). The node head and the depth of the node information are ignored since both values can be derived from Figure 5.8 easily. Note that the routing information in P_4 will be pushed to its descending leaves or nodes. By using the node head, the information pushing is bounded to one level, thus the update complexity can be reduced. Yet the number of leaves is not optimal. Consequently, two improvements, which can further reduce the number of leaves, are presented.

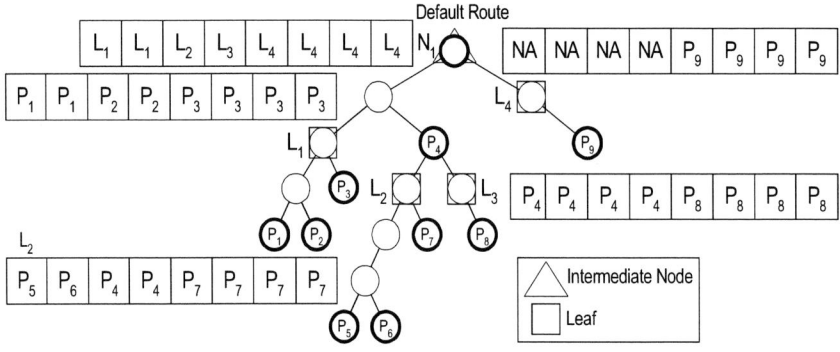

Figure 5.8. Generated LS trie ($N=3$, $M=3$).

Firstly, the algorithm is modified to increase the leaf utilization by raising the leaf position. For example, the L_1, L_2 and L_4 in Figure 5.8 only indicate at most three routing prefixes which is much fewer than the maximum prefixes available in a leaf (2^M). To revise it, a new rule is added to the algorithm. The leaf will be generated only when the difference between the current node and the longest prefix invoked is equal to M. In case the difference is longer than M, the algorithm will be called recursively. As a result, there might be another leaf under a generated leaf, such as two route prefixes which one is a shorter prefix of another one.

Secondly, some prefix information can be included in the intermediate node directly without adding another leaf. For example, L_4 can be erased by recording the prefix P_9 in N_1. The modified construction algorithm is shown as follows and the enhanced forwarding table is shown in Figure 5.9. Obviously, the number of leaf nodes is greatly reduced.

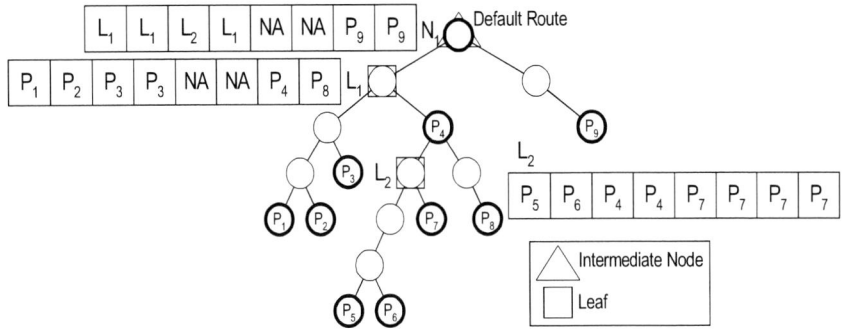

Figure 5.9. Enhanced LS trie ($N=3$, $M=3$).

Modified LS trie Construction Algorithm
Input: The root of the binary tree which is constructed from the routing table.
Output: The forwarding table.
Constructor (Parent, Root, Depth, Prefix) {
  IF ((the deepest child depth in the subtree - Depth) is equal to M)
    Generate_Leaf (Parent, Root);
  ELSE {
   If (Root– >Prefix not equal to NULL)
    Prefix=Root– >Prefix;
   If ((Depth mod N) is equal to 0)
    Parent=Generate_Intermediate_Node(Root,Prefix);
   Constructor(Parent, Root– >Left_Child, Depth+1);
   Constructor(Parent, Root– >Right_Child, Depth+1);
   If ((the length of the longest unprocessed prefix - Depth) is equal M)
    Generate_Leaf(Parent,Root);
  }
}

By using the algorithm, the detailed child information is recorded in the intermediate node to guide the lookup. Also, it is different for those existing schemes to use level compression [12,13]. The idea used in the existing schemes might generate unused nodes, but not in the scheme.

To perform an IP lookup, the destination address is split into N-bit chunks, and these chunks are used to follow a path through the tries until a leaf address or next-hop value in the node pointer is fetched. In the latter situation, the search procedure is terminated. Otherwise, relative depth of the fetched leaf is recorded to decide the M-bit chunk of the address which can be used to access the entry in the leaf. If NA is read in the intermediate node or leaf, for example, to search "100000" in Figure 5.9, the BMP information in the node head would be used.

5.2.2. Route Update

As the topology of the network changes, new routing information is disseminated among the routers, leading to changes in routing tables. As a result of the change, one or more entries must be added, updated or deleted from the table. Because the action of modifying the table can interfere with the process of forwarding packets, various route updates and the relative procedures are considered.

The route updates can be separated into three conditions: **route change**, **route insertion** and **route deletion**. Before considering the route update procedure, the field "prefix length" is necessary to be added to the next-hop field of the leaf in the forwarding table maintained by the network processor in Figure 5.4.

- **Route Change:** For the prefixes located in the leaf, the update can be done by checking and updating the entries with equal length in the leaf. Otherwise, the BMP information of the intermediate node head or its descending leaves will be modified. Thus only the leaves/nodes in one level will be affected in the worst case. As a result, maximum $2^M \times 2^N$ entries will be modified.

  While the head information of the descending nodes is changed, this update will not be forwarded to its leaves/nodes. However, this requires slight change in address lookup procedure. Originally, only the head information in the last node before fetching the leaf will be used. This might cause error result after route changes because the update information may not be broadcasted to the nearest node. To correct this, the longest prefix information in the head is memorized while traversing the trie for possible referring.

- **Route Insertion:** When a route prefix is inserted, the procedure is akin to the route change. However, it might need to generate a long path due to a long prefix insertion. In this case, maximum $32/N$ nodes will be generated.

- **Route Deletion:** The procedure is equal to change the route of the deleted prefix to the route of its shorter prefix. To accomplish this update, the route information of the last matched prefix before reaching the deleted one will be used to refresh the entries which refer to the deleted prefix.

In the worst case, $maximum(2^M \times 2^N, 32/N \times 2^N)$ entries will be updated for IPv4. Assuming that the 5-ns SRAM is used with setting $N = 4, M = 4$, the worst-case update time is 1.28 μsec

5.3. Core-Leaf Decoupling

By observing the data structure of the trie, one may find that most memory accesses refer to the nodes (we use the term *core* to refer the internal nodes of the trie hereafter.) since only one leaf will be accessed in a address lookup. Thus if the core size can be reduced to fit into a higher-speed memory, the performance can be improved dramatically. For example, the core can be located into the CPU L1 cache and the leaves to the CPU L2 cache. If the speed of the L1 cache is twice of the L2 cache, the total lookup time will be reduced to half approximately. By extending the size of leaf, the number of nodes will decrease as well as the core size.

Note that the total required memory space might increase due to a larger-size leaf is used. To alleviate this, a simple array compression scheme can be used. As described in [11], only the variable part in the next-hop array is recorded in a separate entry-array and the original one will be replaced with a bitmap. Each bit in the bitmap maps to an entry in the leaf where the bit is set to one when the array values change. In spite the compression can reduce the size of leaves, one extra memory access is required to access the entry-array. An example of table compression for the L_2 in Figure 5.9 is shown in Figure 5.10.

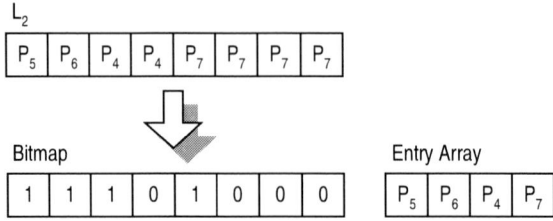

Figure 5.10. Table Compression Example.

5.4. Performance Analysis

In Table 5.1, the performance of the scheme ($N = 4, M = 4$) is presented. The value in the parentheses is the number of memory accesses. Even for the large routing table with more than one hundred thousand entries, the table size is small enough to fit into the high speed SRAM. We assume that the memory access time is 5 ns and nine memory accesses are required to accomplish one lookup, the worst-case lookup time is less than 50 ns, i.e., 20 MPPS can be achieved in the worst case.

Table 5.1. Performance Evaluation with Six Routing Tables.

Site	Routing Prefixes	Memory Required (Kbytes)	Worst-case lookup performance (nsec)
AADS	24,470	373	40
Mae-East	58,101	565	45
Mae-West	36,943	454	45
PacBell	32,338	428	40
Paix	13,395	254	40
NLANR	102,271	866	45

In Table 5.2, we further compare other existing algorithms with the LS tries. The worst-case lookup time of LC tries [13] is not addressed in the literature so we fill it with average lookup time. Obviously, the scheme outperforms the existing schmes in lookup speed. Although we use a table which is much larger than the tables used in Lulea scheme, LC tries and multibit trie, the average required storage is the lowest besides the Lulea scheme. Also, except Patricia trie and multibit trie, the rest of schemes all potentially require changing the complete data structure during a route update, thus feature slow insertion/deletion speed.

Table 5.2. Comparison with Other Existing Works.

Previous Schemes	Worst Case Lookup Time (ns)	Memory Required (Kbytes)
Patricia trie	1,650	3,262
Binary search on hash tables	650	1,600
Lulea scheme	409	160
Multiway search	330	950
LC tries	(*)500	464
Multibit trie	236	640
LS tries	45	866

(*) This is the average performance since the worst-case performance is not addressed.

To test the performance of core-leaf decoupling, the setting of N and M is changed to 4 and 8, respectively. The value in the parentheses is the required storage after applying compression. Obviously, the size of the nodes is much less than that of the leaves. Even with compression, the size of the nodes is about third to quarter of the leaf size. Such as the leaf size of NLANR is compressed from 1.2 Mbytes to 320 Kbytes. Thus the total size of the trie is less than the 512 Kbytes L2 cache of the modern CPUs.

Table 5.3. Performance Evaluation with Core-Leaf Decoupling.

Site	Routing Prefixes	Core Size (Kbytes)	Leaves Size (Kbytes)	Total Size (Kbytes)
AADS	24,470	40	628 (166)	668 (206)
Mae-East	58,101	51	867 (230)	918 (281)
Mae-West	36,943	45	738 (196)	783 (241)
PacBell	32,338	42	699 (185)	741 (227)
Paix	13,395	33	424 (112)	457 (145)
NLANR	102,271	119	1,201 (319)	1,320 (438)

5.5. Summary

In this section, we show how IP-routing tables can be succinctly represented and efficiently searched by structuring them as LS trie. Furthermore, incremental update is supported. Our data structure is simple and not based on any ad hoc assumptions about the distribution of the prefix lengths in routing tables. Even a the routing table with more than 100,000 entries, the size of the forwarding table is still smaller than 900 Kbytes in our approach. The incremental update algorithm is also provided. Each address lookup can be accomplished within nine memory accesses. With the heuristic, core-leaf decoupling, the frequent-access nodes can be put into high speed memory. Through simulation, the scheme outperforms most existing schemes.

Chapter 6

Fast Packet Classification by Using Tuple Reduction

In addition to the basic packet forwarding based on destination address, the new requirements are emerging. To support security, quality of server (**QoS**) or specific business policies, user traffic may be further classified according to a maximum of eight fields: source/destination IP address (32 bits), source/destination transport-layer port numbers (16 bits for TCP and UDP), type-of-service (**TOS**) field (8 bits), protocol field (8 bits) and transport-layer flags (8 bits) with a total of 120 bits. This new forwarding type is called packet classification. Packet classification offers increased flexibility: it gives a router the capability to block traffic from a dangerous external site, to reserve bandwidth for traffic between two specific sites or to give preferential treatment to one kind of traffic over other kinds. Traditional routers do not provide service differentiation because they treat all traffic in the same way. The process of mapping packets to different service classes is referred to as *policy-based routing*. In this section, we present the background of packet classification and a new algorithm in [24].

6.1. Introduction

IP networks are rapidly evolving toward a QoS-enabled and multimedia-friendly infrastructure. In the **DiffServ** model, resources are allocated differently to various aggregated traffic flows according to the differentiated services codepoint (DSCP field). The edge routers positioned at administrative boundaries, for example, the boundary between two ISPs, as shown in Fig. 6.1, set the DSCP field. Packet classification is the process of identifying packets based on specific rules. It has been extensively employed in the Internet for secure filtering and service differentiation by administrators to reflect policies of network operations and resource allocation. Using the pre-defined filters, the packets can be assigned to a given DSCP field. Packet classification is a key component that determines to which forwarding class a packet belongs. Clearly, the performance of packet classification is important in the deployment of new services, including VoIP, VoD and video conferencing, which have strict QoS requirements. The potential future popularity of multimedia services is such that the filter database will be large. Also, packet classification with a potentially large number of filters is difficult and exhibits poor worst-case performance [25].

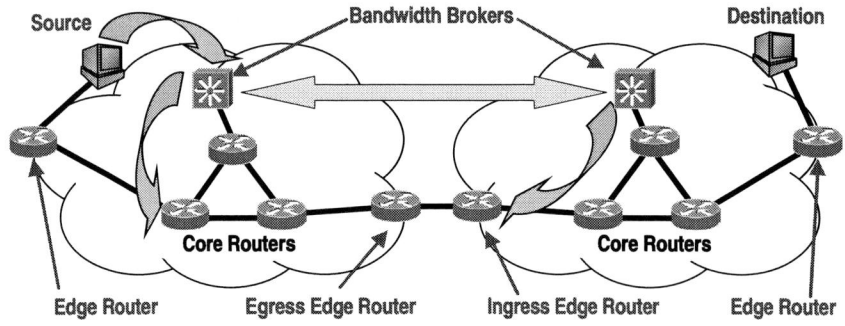

Figure 6.1. DiffServ Framework.

The classifier and the filter must be defined before the problem can be formally described. A classifier includes a set of filters to divide an incoming packet stream into multiple classes. A filter $F = (f[1], f[2], \ldots, f[k])$ is called k-dimensional if it has k fields, of which each $f[i]$ is a variable length prefix bit string, a range or an explicit value of a packet header. A filter can be any combination of fields; the most common fields are the IP source address (SA, 32 bits), the destination address (DA, 32 bits), the protocol type (8 bits), port numbers (16 bits) of source/destination applications and protocol flags in the packet header. A packet P is said to match a particular filter F if for all i, the i_{th} field of the header satisfies $f[i]$. Each filter has an associated action. For example, the filter F=(140.113.*, *, UDP, 1090, *) specifies a rule that addresses a flow belongs to the subnet 140.113, and uses the progressive networks audio (PNA); the rule may assign the flow's packets with high queueing priority. The filter is usually assigned a cost to define its priority of the matched filters. The least-cost matched filter will act to process the arriving packets, such that the packet classification problem is a least-cost problem.

Hashing is an extensively applied method for performing fast lookup. Many hash-based schemes have been presented for solving the Internet lookup problem [14, 26, 27]. The tuple space search is a well-known two-dimensional (*source address prefix, destination address prefix*) solution which is based on multiple hash accesses for various filter-length combinations [26, 28]. The rectangle search, one of the tuple-based algorithms, is proposed and shows the lower bound $O(2W - 1)$ of the lookup speed, where W is the length of the IP address. The rectangle search is highly scalable with respect to the number of filters. However, it suffers from the memory-explosion problem. Also, the lookup is not sufficiently fast. For example, through experiments, the required entries are determined to be as large as twelve-fold filters and each packet classification requires approximately 40 hash accesses in a 100K-filter database. Consequently, it cannot support a gigabit throughput.

In the following, an efficient scheme for improving the rectangle search is presented. The scheme consists of two parts. The first presents the **"Tuple Reduction Algorithm"**. Duplicating the filters dramatically reduces the number of tuples. Dynamic programming is used to optimize the tuple reduction and two heuristic optimization methods are introduced to simplify the optimization process. The primary goal of the spatial optimization approach is used to minimize the required storage while the speed optimization approach can further increase the lookup speed. By the tuple reduction, all the performance metrics, including storage, speed and implementation complexity, are enhanced. The second part presents the **"Look-ahead Caching"** scheme, which can further improve the lookup performance. The basic idea is to prevent unnecessary tuple probing by filtering out the **"un-matched"** situation of the incoming packet. Obtained by combining the tuple reduction algorithm and look-ahead caching, the experimental results show that the lookup speed increased by a factor of six while requiring only around one third of the required storage.

6.2. Previous Works

Several algorithms for classifying packets have recently appeared in the literature [25, 26, 29–34]. They can be grouped into the following classes: linear search/caching, hardware-based solutions, grid of tries/cross-producting, recursive-flow classification, and hash-based solutions. The following briefly describes the important properties of these algorithms. Assume that N is the number of the filters, k is the number of classified fields and W is the length of the IP address.

Linear Search/Caching: The simplest method for packet classification involves a linear search of all the filters. The spatial and temporal complexity is $O(N)$. Caching is a technique frequently used at either the hardware or the software level to improve the performance of linear search. However, the performance of the caching depends critically on each flow's having large number of packets. Also, if the number of simultaneous flows exceeds the cache size, then the performance is severely degraded.

Hardware-based Solutions: A high degree of parallelism can be implemented in hardware to provide a speed-up advantage. In particular, ternary content addressable memories (TCAMs) can be used effectively to look up filters. However, TCAMs with a particular word width cannot be used when flexibility of the filter specification is required. Manufacturing TCAMs with sufficiently wide words to contain all bits in a filter is difficult. It also suffers from the problem of power consumption and scalability [28]. Lakshamn et al. presented another scheme that depend on a very wide memory bus [30]. The algorithm reads Nk bits from memory, corresponding to the BMPs in each field, and determines their intersection to find a set of matching filters. The memory requirement for this scheme is $O(N^2)$. The hardware-oriented schemes rely on heavy parallelism, and involve considerable hardware cost; the flexibility and scalability of hardware solutions is very limited.

Grid of Tries/Cross-producting: Specifically for the case of two-field filters, Srinivasan et al. [30] presented a trie-based algorithm. The algorithm has a memory requirement of $O(NW)$ and requires $2W - 1$ memory accesses per filter lookup. It is a general mechanism, called cross-producting, which involves BMP lookups on individual fields, and the use of a pre-computed table to combine the results of individual prefix lookups. However, this scheme suffers from a $O(N^k)$ memory blowup for k-field filters, including $k = 2$ field filters.

Recursive-flow Classification: Gupta et al. presented an algorithm that can be considered to be a generalization of cross-producting [25]. After BMP lookup is performed, a recursive flow classification algorithm hierarchically performs cross-producting. Thus k BMP lookups and $k - 1$ additional memory accesses are required per filter lookup. The algorithm is expected significantly to improve average throughput, but it requires $O(N^k)$ memory in the worst case.

Also, in the case of two-field filters, this scheme is identical to cross-producting, and hence has a memory requirement of $O(N^2)$.

Hash-based Solution: This solution is motivated by the observation that, although filter databases include several different prefixes or ranges, the distinct prefix lengths tend to be few [26]. For example, backbone routers have around 60K destination address prefixes, but only 32 distinct prefix lengths exist. Hence, all the prefixes can be divided into 32 groups, one for each length (W). Since all prefixes in a group have the same length, the prefix bit string can be used as a hash key, leading to a simple IP lookup scheme, which requires $O(W)$ hash lookups, independent of the number of prefixes. The algorithm of Waldvogel [14] performs a binary search over the W length groups, and has a worst-case time complexity $O(logW)$.

The tuple space idea generalizes the foregoing approach [14] to multi-dimensional filters [26]. A tuple is a set of filters with specific prefix lengths, and the resulting set of tuples is called a **"tuple space"**. Since each tuple has a specific bit-length of each field, these bit-lengths can be concatenated to create a hash key, which can be used in performing the tuple lookup. The matched filter can be found by probing each tuple alternately, and tracking the least-cost filter. For example, the two-dimensional filters $F = (10*, 110*)$ and $G = (11*, 001*)$ both belong to the tuple $T_{2,3}$ in the second row and third column in the tuple space. When searching for $T_{2,3}$, a hash key is constructed by concatenating two bits of the source field with three bits of the destination field. Even a linear search of the tuple space represent a considerable improvement over a linear search of the filters since the number of tuples is typically much smaller than the number of filters.

The rectangle search, a tuple-based algorithm, was proposed to improve the performance of the tuple lookup [26]. The lower bound has been demonstrated to be $O(2W - 1)$ given a $O(W \times W)$ rectangular tuple space, where W is the number of distinct prefix lengths. The primary aim is to eliminate a set of tuples during each probing, as depicted in Fig. 6.2. Tuples above T are eliminated if the probe of tuple T returns **"Match"**. Otherwise, tuples to the right of tuple T are discarded. Markers and a pre-computation mechanism are required to reach this goal. Assuming that the number of filters is N, a rectangle search requires NW memory space. Notably, reducing the number of distinct lengths would increase lookup performance. Furthermore, the number of generated markers can be eliminated under certain conditions.

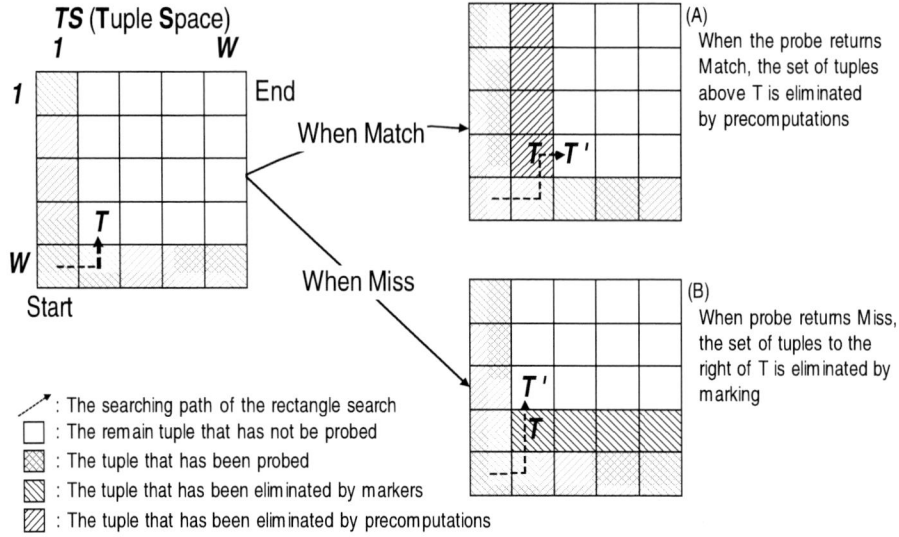

Figure 6.2. Rectangle Search Algorithm.

6.3. Tuple Reduction Algorithm

For a given **TS**, pre-computation and markers are used to perform a rectangle search. Assume that the filter database contains 203 filters, which occupy five tuples, as shown in Fig. 6.3. Each filter must generate one marker in each left-side tuple to indicate the existence of a long potential matched filter. Accordingly, 703 markers are generated, resulting in 906 entries. The size explosion is caused by the unbalanced distribution of routing prefix lengths, which is currently common in the Internet [10, 12–14]. Clearly, reducing the number of tuples to promote a native rectangle search is an important issue.

6.3.1. Filter Expansion

The lookup performance of the rectangle search can be improved by reducing the distinct lengths. A well-known capability, based on duplicating entries, can be used, as described in [12]. Dynamic programming is used to minimize the increasing storage required by the duplication of entries. The other algorithms [9–11, 13] are based on a similar mechanism, but have additional compression schemes. Briefly, the algorithms in this category depend on either large

Fast Packet Classification by Using Tuple Reduction 59

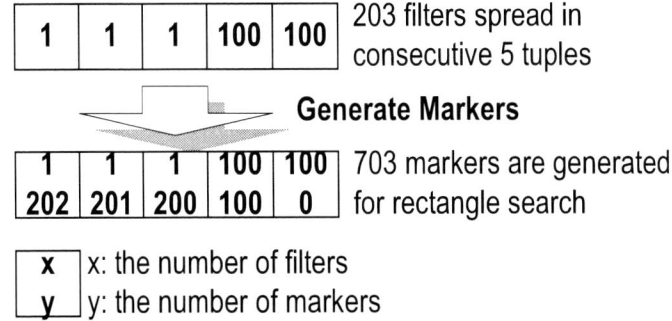

Figure 6.3. Relationship between the Filters and the Markers.

storage or complex compression logic, which must be traded off to increase throughput. However, faster lookup can be obtained with a smaller storage in the rectangle search because the reduction of the tuple also results in fewer generated markers. Figure 6.3 considers the same example as presented in Fig. 6.4 to facilitate the explanation. The three filters with short prefixes are expanded to the fourth tuple, corresponding to a longer filter. The number of required markers is reduced drastically by the expansion of filters.

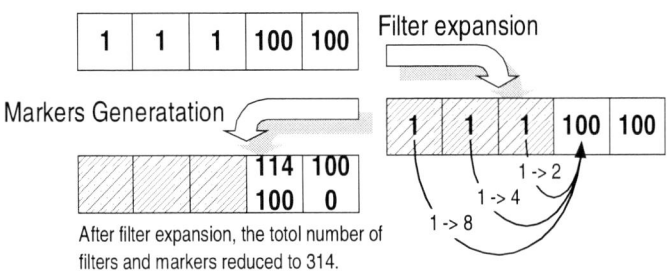

Figure 6.4. Tuple Space with Filter Expansion.

Figure 6.5 depicts a two-dimensional filter expansion. The filter $f(11*, 10*)$ at the tuple $T_{2,2}$ is expanded to the destination tuple $T_{3,3}$. Firstly, the source prefix of f is expanded from $11*$ to the prefixes $110*$ and $111*$. By the same procedure, the destination prefix is expanded from $10*$ to $100*$ and $101*$. After both the source and destination prefixes are expanded, the cross-product of the two sets of prefixes are obtained, yielding four new filters $f_1(110*, 100*)$, $f_2(111*, 100*)$, $f_3(110*, 101*)$ and $f_4(111*, 101*)$, whose actions equal that of

f.

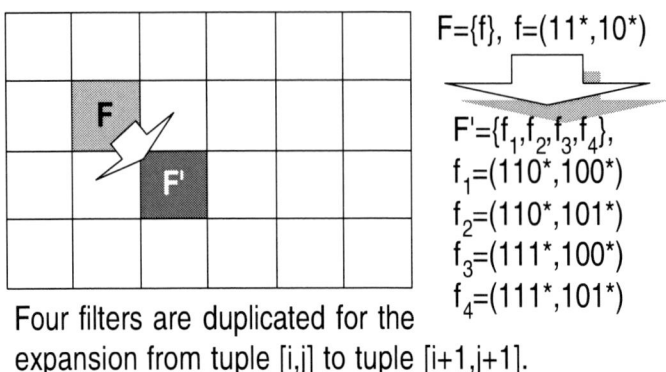

Figure 6.5. Filter Expansion.

6.3.2. Two-Dimensional Tuple Reduction to Minimize Storage

The tuple reduction algorithm presented herein is a space-optimization scheme in which the decision to expand the filter is made independently for each tuple. The design principle is to minimize the required storage by converging the tuple space. In the following, dynamic programming and two heuristic methods are introduced to minimize storage.

Dynamic programming is used to perform the optimal filter expansion with minimal cost. Figure 6.6 illustrates the cost of a filter expansion. The cost is determined by several contributing factors, including the number of duplicated filters due to the expansion as described above, the number of original markers in the source tuple and the number of markers generated of the duplicated filters. The original markers are eliminated by the expansion of the filter. The number of markers of the duplicated filters is also important. The original filters, the duplicate filters and the markers must be simultaneously considered to determine the minimal expansion cost. The minimal expansion cost can be formulated as a recursive equation, as shown below. The number of the filters (tuples) in the tuple (tuple space) is defined as $|T|$ ($|TS|$). To simplify the expression, a tuple at the i_{th} row and the j_{th} column is labeled as $T_{M(i,j)}$ using the following one-to-one mapping function $M(i,j) = (i+j-2) \times (i+j-1)/2 + j$. Consequently, the tuple space can be equivalently represented by a set $T_1, T_2, \ldots, T_{|TS|}$.

$$DynaP(TS) = \min_{i=1}^{|TS|}\{DynaP(TS-T_i) + \min_{j=1}^{|TS|}\{Expansion(T_i,T_j)$$
$$-Sum\_of\_Filters(T_i)\}\}, where \qquad (6.1)$$

$$Expansion(T_{src},T_{dest}) = 2^{Dist(T_{src},T_{dest})}$$
$$\times |T_{src}| \times Left\_Tuples(T_{dest}); \qquad (6.2)$$

$$Sum\_Of\_Filters(T_{M(R,C)}) = \sum_{col=C}^{C\leq W} |T_{M(R,col)}|; \qquad (6.3)$$

$$Left\_Tuples(T) = \text{Number of tuples in the left side of } T. (6.4)$$

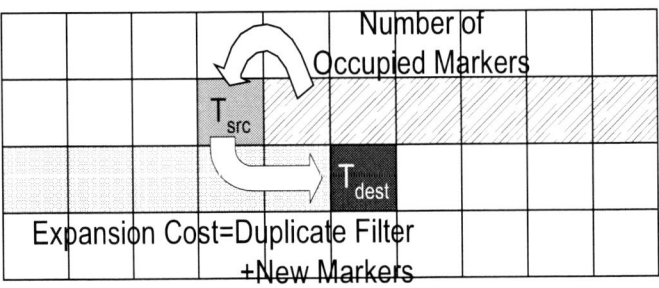

Figure 6.6. Dynamic Programming for Expanding the Filter.

The cost function is $f(\alpha) = \alpha \times f(\alpha-1) + \alpha^2$, where α is the number of tuples and α^2 is the complexity of determining to where a tuple is expanded. Thus, the cost-function is exponential ($f(\alpha) \leq 2^\alpha$). Although an optimal solution can be derived, dynamic programming is time-consuming.

6.3.3. Semi-Optimization Algorithm

A heuristic semi-optimization algorithm that restricts the expansions is adopted to reduce the time complexity. The new algorithm expands tuples only to those with large number of filters because the cost of expansion to those tuples is usually higher than that to others. Thus, expansion begins from tuples with the fewest filters. The number of expansion combinations can be considerably reduced to $|TS| \times |TS|$. Moreover, expansion is allowed within a predefined region, such as a 3×3 region, from the source tuple, $T_{i,j}$ to $T_{i+2,j+2}$. The semi-optimization algorithm is described below. The tuples are sorted by the number of filters. In the iteration i ($1 \leq i \leq |TS|$), *Cost_Old* refers to the current cost of T_i without expansion. T_{dest} is the best expanded tuple of tuple T_i within the restricted region. *Cost_New* shows the new cost of tuple T_i is expanded to T_{dest}. If the *Cost_New* is greater, it continues to process the subsequent iteration. Otherwise, the filters in T_i are expanded to T_{dest}.

Semi-Optimization Algorithm
[Functions]
Markers(T) : Generated markers for the filters in tuple T.
Optimal_Expansion(T) : Optimal expansion for tuple T within the restricted region.
[Input] Ascending-ordered Tuples $\{T_1, T_2, \ldots, T_M\}$
[Output] Optimized Tuples
For ($i = 1; i \leq M; i++$) {
        Cost_Old $= |T_i| + $ *Markers*$(T_i) + $ *Sum_of_Filters*(T_i);
        $T_{dest} = $ *Optimal_Expansion*(T_i);
        Cost_New $= $ *Expansion*(T_i, T_{dest});
        If (*Cost_Old* $\geq$ *Cost_New*)
                Expand Tuple T_i to T_{dest};
}

6.3.4. Tuple Reduction in Area of Sparse Filter

The semi-optimization algorithm achieved a near-optimum rectangle search. Some tuples are eliminated by trading off storage. Most existing IP address lookup schemes have to deal with an unbalanced length-distribution of routing prefixes. Although large filter databases are not yet commonly available, the authors believe that unbalanced distributions will continue to be prevalent. A randomly generated filter database with 100K filters is used to illustrate the phenomenon, as shown in Fig. 6.7. Each tuple is positioned according to its length combination and colored according to the number of filters. Most filters are in the region from the upper-left tuple $T_{16,16}$ to the bottom-right tuple $T_{24,24}$ because most routing prefixes have a length of 16-24 bits [14]. Relatively few tuples are occupied by filters in regions **A, B, C, D** and **G** when the **TS** is separated into nine regions. Although regions **F, H** and **I** include more occupied tuples, the density of filters is lower than that of tuples in region **E**. Essentially, too many tuples with few filters represent the primary factor reducing the search rate because the cost of access is the same, independently of the number of filters in each tuple.

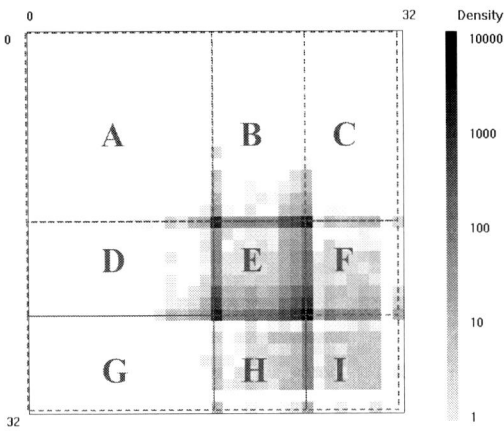

According to the density of the filters, we can separate the tuple space into 9 regions.

Figure 6.7. Unbalanced Filter Distribution.

This work proposes an enhancement of tuple reduction by collecting filters within sparse regions into a target tuple in the lower-right hand corner of the region. Thus, the nearest non-empty tuple is expanded into the target tuple. If the number of aggregated filters does not exceed a certain threshold, $MAX\_FILTERS$, the procedure will be repeated. Otherwise, another target tuple in this region is chosen and the procedure is then performed.

6.3.5. Further Optimizing of Lookup Speed

The optimization algorithms presented above efficiently reduce the required storage; however, the lookup speed is not yet maximal, primarily because the self-absorbed expansion of each tuple cannot systematically reduce the number of rows and columns. Hence, the other heuristic approach for eliminating the tuple probes by restricting the filter expansion to specific rows/columns is considered. Restated, the selected rows and columns interlace to generate a set of grid-cells and the filters are expanded to the nearest tuples in the grids.

The required storage will be enlarged because of the restriction, and dynamic programming is used to minimize it.

Several items must first be defined. Assume that the resulting tuple space has r rows and c columns. The function $F(P)$ is defined as the minimal cost of a specific (r,c) combination where $r+c = P$ and $2 \leq P \leq 2W$, as shown in Eq. 6.5. For each (r,c) combination, $\binom{W}{r} \times \binom{W}{c}$ row/column selections exist. The closest tuple to T_i is expressed as $T_{closest}$, which conforms to the (row, column) selection results. The resulting (row, column) selection will satisfy $\min_{P=1}^{2W} F(P)$. However, the complexity of computation is markedly increased $(2W \times W \times \binom{W}{r} \times \binom{W}{c})$ since the results of each combination are calculated from scratch. Although recording the executed results for later reference can reduce costs, a more efficient approach is introduced here. First, the number of filters of each row and column are calculated. Thus, the rows and columns are sorted according to the their summations. The minimal cost of each (r,c) combination can be derived easily by selecting the rows and columns with the highest value. Although the results may not be optimal, the computational cost is reduced greatly.

$$F(P) = \min_{\substack{for\ all\ (row,column) \\ combinations}} \sum_{i=1}^{|TS|} (Expansion(T_i, T_{closest}) - Sum\_of\_Filters(T_i)), \quad (6.5)$$

$$\text{where } r+c = P.$$

6.4. Look-Ahead Caching Mechanism

The space occupied by each filter can be intuitively assumed to be small since only specific services require the setting up of filters for packet classification. Also, multiple filters are unlikely to be defined over a specific address space. Thus, filter matching might be rare in tuple probing. Look-ahead caching is presented to filter out the **"un-matched"** case of each incoming packet using dual-hash architecture to eliminate unnecessary tuple accesses during the rectangle search.

The look-ahead cache includes as many tables as tuples. Figure 6.8 depicts the basic mechanism. For each incoming packet, the look-ahead cache is probed firstly. If the probing indicates a miss, no matched filter is present in the corresponding tuples; thus, the tuple access can be bypassed. Otherwise, the tuple will be probed for **"possible"** matched filter.

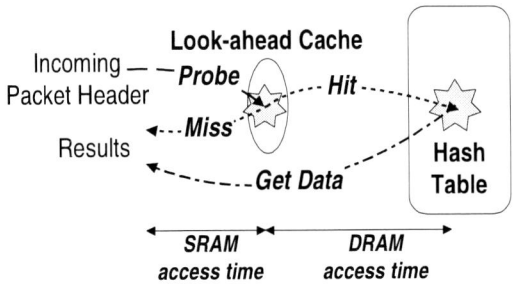

Figure 6.8. Concept of the Look-ahead Caching.

For each table, the number of entries is identical to the number of corresponding tuples but each entry is much smaller so that they can fit into high-speed SRAM. An entry in the look-ahead cache has three fields, **Tag**, **Empty** and **Collision**, as shown in Fig. 6.9. The source and destination prefixes are extracted from the original filter and mapped onto the Tag using the function $TAG\_FCN()$. The Tag is used to specify whether a filter occupies the entry. Its length is defined by the size of SRAM. The **E** (Empty) bit states the status of the corresponding entry in the tuple. If this entry contains no filter (that is, the E bit is set to one), the look-ahead cache probing returns "false" and tuple probing can be skipped. The **C** (Collision) bit states whether multiple filters map onto the entry (such that, the collision occurs in the hash entry).

Figure 6.10 presents the tuple space search with look-ahead caching. When a tuple is initially probed for a packet with header Y, the entry location in the look-ahead cache is calculated using the hash function $INDEX\_FCN(Y)$ and the tag value $TAG\_FCN(Y)$. After the cache entry is retrieved, the **Empty** bit and the **Collision** bit of the entry are tested. If the **Empty** bit of the entry is set to one, then the entry is empty, and a **miss** is returned. Otherwise, the **Collision** bit is checked. If it is set, the collision occurs, and these filters must be accessed and compared one by one. Otherwise, $TAG\_FCN(Y)$ is compared with the Tag of the fetched entry. If they differ, a **miss** is returned, and tuple access can thus be avoided. Otherwise, the corresponding entry in the tuple is probed for

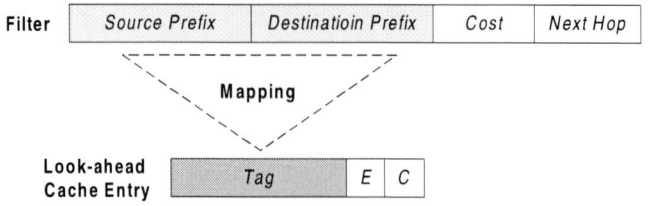

Figure 6.9. Data Structure of the Look-ahead Cache Entry.

possible matching.

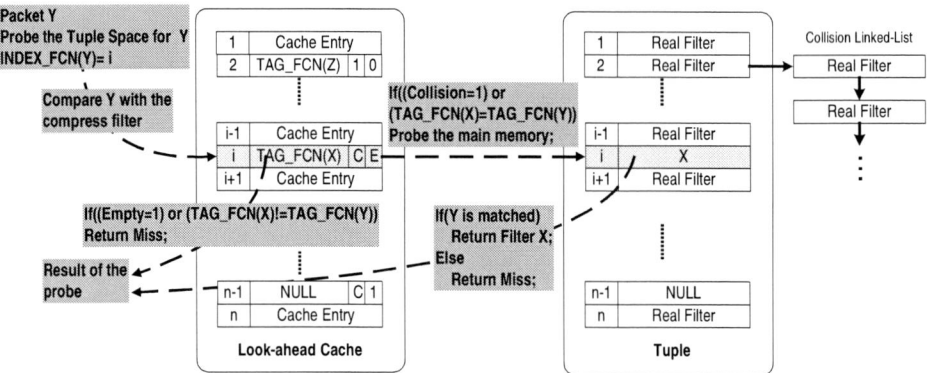

Figure 6.10. Lookup Procedure with Look-ahead Caching.

6.4.1. Realization of Parallel Hardware

Experiments that will be discussed in Section 6.5. will reveal that look-ahead caching can markedly enhance the lookup performance of a rectangle search. However, the look-ahead cache always waits for the access to a tuple since the access time of DRAM is much greater than that of SRAM. Thus, the design principle that governs hardware implementation is to keep the look-ahead cache as busy as possible.

Fast Packet Classification by Using Tuple Reduction 67

According to Fig. 6.11, the implementation involves a network processor, SRAM and DRAM. A network processor includes multiple processing elements (PEs) and request queues. PEs are connected to request queues through the shared buses. The look-ahead cache and tuples are located in the SRAM and DRAM, respectively. The request queues are used to resolve the memory bottleneck by simultaneously accessing the look-ahead cache and the tuples. When the queue of the look-ahead cache is empty, the idle PE issues another packet header for classification. Sets of PEs ($\lceil$DRAM access time/SRAM access time$\rceil$) are required to keep the look-ahead cache busy. Let the access time of the DRAM and SRAM be 25 ns and 5 ns, respectively. $\lceil 25 \text{ ns}/5 \text{ ns} \rceil = 5$ PEs are deployed. The look-ahead cache is always busy so the mean lookup time is reduced to *"Number of Tuples Probing×DRAM Access Time"*. However, deploying multiple PEs raises a new issue. That is, in the worst case, the time taken to accomplish single packet classification is the product of the number of PEs and "the worst case lookup time without parallel implementation". The packet might undergo a longer delay in the router than it would under the native scheme. For example, each classification is assumed to require P tuple accesses. Given five PEs, the maximum delay in the classifier is $5 \times (P \times 25 \text{ ns}) = 125P$ ns. In the worst-case of 64 tuple probes, the holding time is 8 μs. However, the delay is tolerable for presently available services.

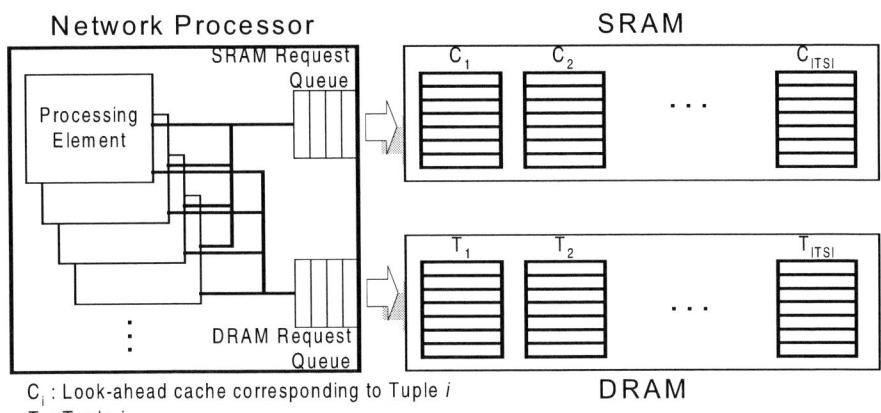

C_i : Look-ahead cache corresponding to Tuple i
T_i : Tuple i

Figure 6.11. Realization using Parallel Hardware.

6.5. Performance Evaluation

This section presents the experiment results to demonstrate that the tuple reduction algorithm increases the lookup speed and reduces the memory requirement. Synthetic filter databases are used to evaluate the performance, mainly because filter databases are normally considered to be commercial secrets and most are relatively small, including those considered in [26] and [25].

The routing table includes 102,309 prefixes, downloaded from the NLANR [8] as a basis for synthesizing filters. Sampling the routing prefixes randomly produces 12 ⟨source prefix, destination prefix⟩ filter databases. The minimal database contains 1K filters; and the sizes of the databases increase to 5k, 10k, 20k, 30K, and so on, and up to 100K filters. Look-ahead caching is not used in the first part of experiments to realize the effect of tuple reduction.

The effect of tuple reduction is firstly specified using the filter distribution, as shown in Fig. 6.12. Figure 6.12(a) shows the filter distribution in the original **TS** for the 100K-filter database. It is comprised of 350 tuples, and leads to several tuples with few filters. After the semi-optimization algorithm is applied, the TS is reduced to 85 tuples, as shown in Fig. 6.12(b). However, several tuples contain few filters (< 100 filters). Applying the tuple reduction to the sparse region (*MAX_FILTERS=1,000*) can reduce the number of the tuples to 55, as shown in Fig. 6.12(c).

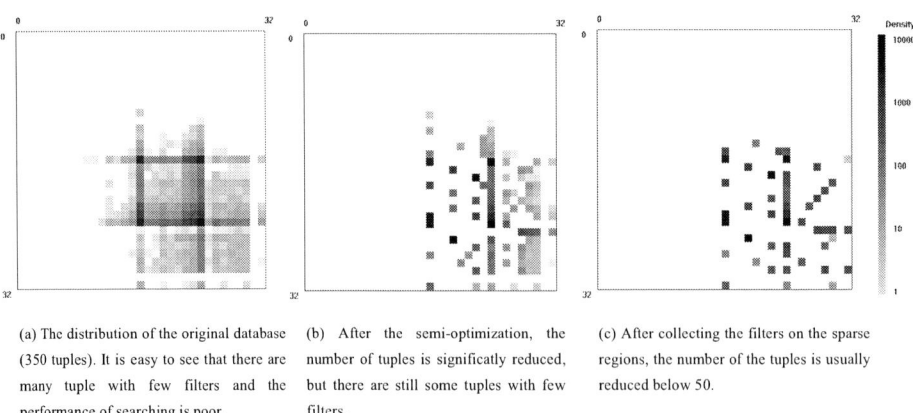

(a) The distribution of the original database (350 tuples). It is easy to see that there are many tuple with few filters and the performance of searching is poor.

(b) After the semi-optimization, the number of tuples is significantly reduced, but there are still some tuples with few filters.

(c) After collecting the filters on the sparse regions, the number of the tuples is usually reduced below 50.

Figure 6.12. Filter Distribution after Filter Expansion.

Figure 6.12(c) clearly shows that expanded tuples are disordered and widespread, resulting in increased search time since it ties to the sum of the numbers of rows and columns. The speed-optimization algorithm can improve the performance by expanding simultaneously the tuples in identical rows or columns. Although the number of occupied tuples is not markedly changed, the numbers of the rows and columns are reduced, as shown in Fig. 6.13.

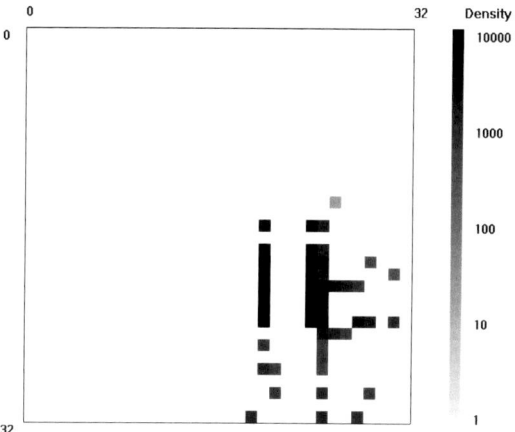

Figure 6.13. Filter Distribution after Speed is Optimized.

Figure 6.14 compares the number of the occupied tuples within various databases. The term "space-optimization" is used to refer to the combination of the semi-optimization algorithm and sparse-region reduction. The curve generated by the rectangle search is varies slightly because of the random generation of the filter databases. Both algorithms outperform the rectangle search since occupied tuples are noticeably eliminated when the optimization of speed further eliminates tuples because of the restrictions of expansion. After tuple reduction, the new schemes can achieve a throughput comparable to that of the rectangle search, even using a simple linear search. Furthermore, the number of occupied tuples is very stable with respect to the number of filters. Consequently, the new algorithms can provide much sustainable throughput. Also, memory management can be simplified with many fewer tuples.

Figure 6.15 presents the number of filters and generated markers. Though the number of entries is proportional to the number of filters, the rate of increase in the number of filters and markers of the new algorithms is much less than that of rectangle search, mainly because tuple reduction also leads to a requirement

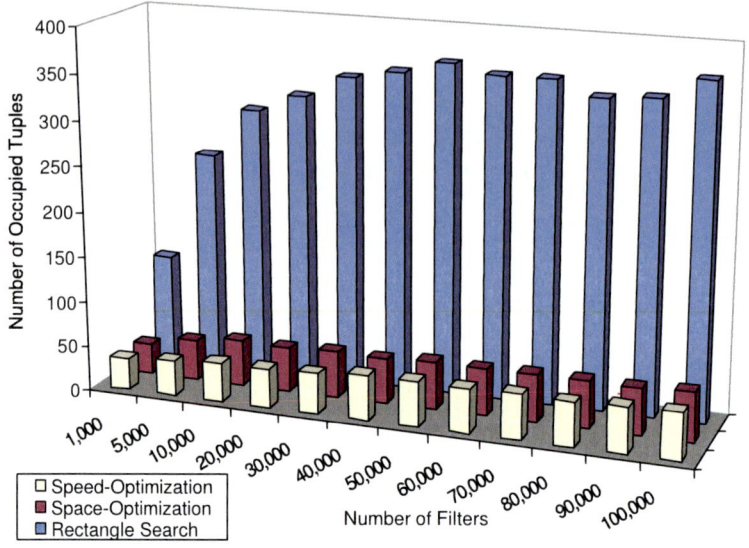

Figure 6.14. Number of Occupied Tuples.

for fewer entries. Clearly, both algorithms are better than the native rectangle search. The speed-optimization algorithm requires more entries than the space-optimization algorithm as a trade-off for fewer tuples.

Table 6.1 lists detailed results concerning occupied tuples and generated entries. The speed-optimization algorithm depends on fewest tuples while the space-optimization generates fewest entries. If an 80-bit filter (32-bit source prefix, 32-bit destination prefix, 8-bit cost and 8-bits nexthop) is used, then the required memory can be derived. Using a 100K-filter database, the three schemes require 12,586 Kbytes, 3,280 Kbytes and 4,909 Kbytes. Apparently, the new schemes are very efficient, and are thus suited to large filter database and implementation using high-speed memory.

One hundred million packets are randomly generated and the number of tuple probes recorded to demonstrate the lookup speed of the new scheme. Figure 6.16 shows the search performance. The number of tuple probes equals the number of occupied tuples since each tuple is probed in a linear search. The scalability of the rectangle-based algorithms is demonstrated. The lookup time is relatively stable with respect to the number of filters. Although the rectangle search significantly improves the linear search, the tuple reduction algorithms can further reduce the tuple probing to a near-optimal level.

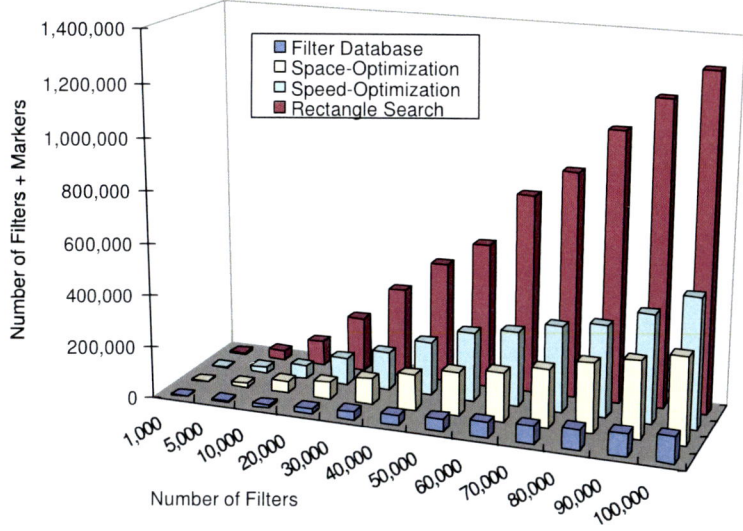

Figure 6.15. Number of Searchable Entries.

Table 6.2 presents in detail the results of the investigations into the search speed. The space-optimization algorithm probes 19 tuples for the least-cost filter in a 100K-filter database, while the rectangle search probes 44 tuples on average. The speed-optimization algorithm can further reduce the number of probes to 14 and triple the throughput of the rectangle search. Simultaneously considering the storage and speed enhancements, the space-optimization and speed optimization improve the rectangle search by factors of 8.8 $[(1,288,823/335,882) \times (44/19)]$ and 8.0 $[(1,288,823/502,734) \times (44/14)]$, respectively. The speed-optimization algorithm can perform one lookup per 575 ns in the worst case when commercially available 25-ns DRAM is used. The average lookup time is 350 ns; that is, two OC-48 links can be supported by an average packet size of 256 bytes.

The effect of the look-ahead cache is now examined. The performance obtained using the rectangle search can be examined since the look-ahead cache can also be coupled to any tuple-based algorithm. The experimental results in Fig. 6.17 indicate that the number of tuple probes in the three schemes is decreased as the size of Tag increases. Even using a zero-length tag, the required tuple probing can be further reduced by the effect of the **E** bit. However, once the size of the tag exceeds 8 bits, the benefit of look-ahead caching is elimi-

Table 6.1. Comparing Storage.

Filter Databases	Rectangle Search		Space-Optimization		Speed-Optimization	
	Tuples	Entries	Tuples	Entries	Tuples	Entries
1K	119	6,472	34	2,775	35	3,572
5K	241	35,440	44	17,110	39	23,566
10K	296	95,206	51	46,616	43	50,971
20K	316	209,574	49	63,805	43	104,224
30K	340	344,592	51	100,260	45	147,224
40K	350	466,714	50	136,367	50	206,482
50K	363	560,438	53	168,339	50	265,299
60K	354	767,865	52	188,627	49	290,059
70K	355	872,355	53	225,106	50	332,293
80K	339	1,043,683	53	270,266	49	355,983
90K	343	1,176,061	52	299,550	51	418,469
100K	366	1,288,823	55	335,882	52	502,734

nated. Two factors may impact the performance. The first is the number of markers. The markers related to each matched filter are probed in the packet classification. Thus, the minimum number of probes is bounded by the number of columns. Another factor is the hashing efficiency. In the experiments, the hash function caused around 26% of the collisions, implying that 26% of cache accesses result in tuple probing, independently of the tag field. Therefore, if the hash efficiency can be improved, the effect of the look-ahead caching may increase.

Table 6.3 presents the detailed results concerning look-ahead caching. Notably, when the tag size falls to zero, the length of the tag is one since 1 bit remains as an **E** bit. In the rectangle search and space-optimization, an 8-bit tag is required to ensure significant performance enhancement. However, the speed-optimization algorithm only requires 2 bits, mainly because disordered tuples in the rectangle search and space-optimization cause pre-computation information and markers to be spread across more tuples, increasing the hit probability of the look-ahead cache. Unlike the tuple storage, the cache size required by space-optimization and speed-optimization is sufficiently small to fit into a high-speed SRAM at relatively low cost.

The following formula is used to calculate the total search time of the tuple reduction schemes.

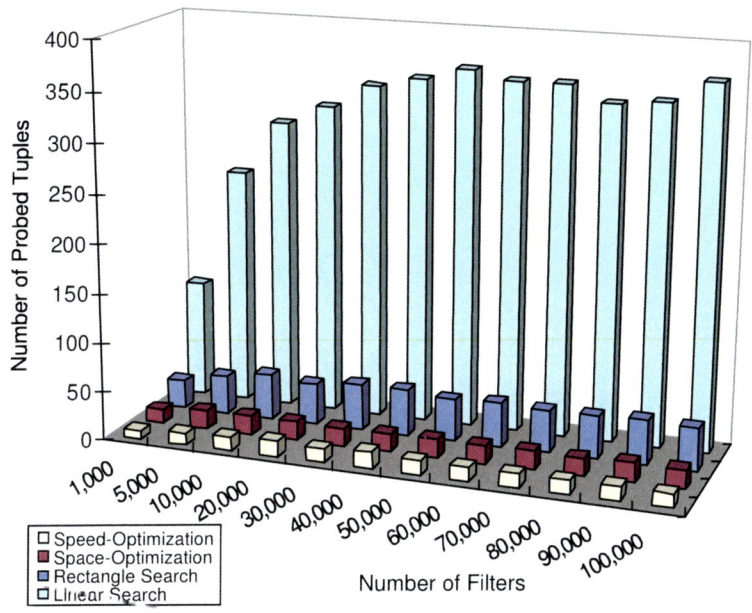

Figure 6.16. Average Search Performance.

Total Search Time = Number of the Look-ahead Cache Probing × SRAM Access Time

+ Number of Tuples Probing × DRAM Access Time

Assume that 5-ns SRAM and an 8-bit tag are used. The mean search time of 270 ns (19 × 5 ns + 7 × 25 ns) can be derived for the space-optimization scheme using 409-Kbyte SRAM. Compared to the rectangle search, which requires 620 ns (44 × 5 ns + 16 × 25), the tuple reduction scheme provides a twofold improvement while requiring only one quarter of the SRAM. Furthermore, deploying pipeline hardware can further enhance these improvements. The maximum packet delay under parallel implementation is 2.375 μs (125 × 19 ns), which is satisfactory. However, the mean lookup time is reduced to 175 ns and an OC-192 throughput is achieved. The mean lookup time of the speed-optimization scheme is the same as that of space-optimization, but using only 245-Kbyte of SRAM.

Table 6.2. Comparing Speed.

Schemes	Rectangle Search			Space-Optimization			Speed-Optimization		
Speed	Min	Avg	Max	Min	Avg	Max	Min	Avg	Max
1K	23	29	36	10	14	21	7	9	17
5K	31	39	48	15	19	24	9	12	20
10K	33	46	57	16	19	28	10	14	22
20K	30	42	53	14	19	27	11	16	22
30K	35	47	58	13	18	26	11	14	22
40K	30	47	59	14	18	26	12	17	24
50K	27	43	56	14	19	27	11	15	22
60K	31	45	60	13	18	26	10	14	21
70K	28	43	56	13	19	27	11	14	22
80K	30	44	57	14	18	27	10	14	22
90K	31	46	59	16	20	29	11	15	22
100K	28	44	59	13	19	26	9	14	23

Table 6.3. Look-ahead Cache Performance.

Schemes	Performance Metrics	Tag Size (bits)					
		0	2	4	8	16	24	32
Rectangle Search	Cache (Kbytes)	157	628	942	1,570	2,826	4,083	5,339
	Average Probes	35	34	34	16	15	15	15
Space-Optimization	Cache (Kbytes)	41	164	246	409	737	1,064	1,391
	Average Probes	15	15	15	7	6	6	6
Speed-Optimization	Cache (Kbytes)	61	245	367	612	1,102	1,592	2,082
	Average Probes	9	7	7	7	6	6	6

6.6. Summary

Packet classification is a highly effective primitive for associating a policy-defined context with each incoming packet, so as to permit packet handling using DiffServ actions to activate multimedia services. However, implementing at high-speeds with a large number of rules is difficult. This study proposes various methods for improving both performance and the storage of the rectangle search algorithm. The way in which filter expansion can be applied to the

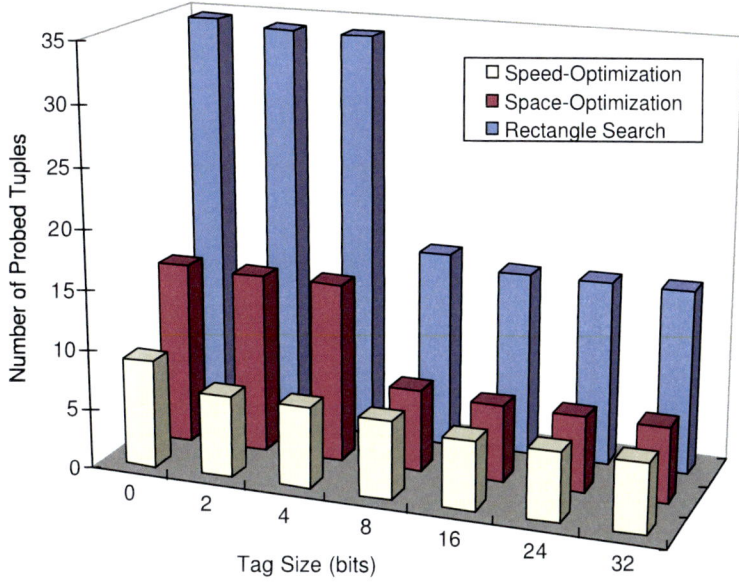

Figure 6.17. Average Number of Tuple Probes.

rectangle search to reduce the number of markers and tuples in the first part is addressed; then space and speed optimization techniques are introduced. Space optimization can minimize storage requirements. A semi-optimization algorithm is also presented by restricting the expansions because of the high cost of dynamic programming. The tuple reduction over a sparse area is introduced to reduce the number of tuples. Therefore, the row and column-based expansion is presented as speed optimization. While the algorithm is not storage-optimized, the lookup time is reduced by the regular tuple distribution. Furthermore, "look-ahead caching", which is based on the assumption of a low filter-hit-ratio in the Internet, is introduced. The experimental results obtained by combining the tuple reduction algorithm and look-ahead caching show that the lookup speed was increased by a factor of six while required storage was reduced to only about a third. With an average packet size of 256 bytes, the tuple reduction scheme can provide OC-192 throughput, even in a large database with 100K filers.

Chapter 7

Conclusion

This chapter presents several efficient and scalable algorithms for IP forwarding engine in Internet routers. These algorithms allow lookups at gigabit speeds even in software. A good lookup algorithm should satisfy: **simplicity, small storage, fast update, IPv6 scalability and parallelism**. However, some of these item might need balance among themselves. For example, a fineness compression may largely reduce the required storage, but might result in increasing complexity and slower lookup speed. On the other hand, with less required storage, if the table can be fitted into the high-speed memory, it might result in higher lookup speed. But the compressed data structure may not support incremental update.

We now review the main ideas of these algorithms and outline the principles underlying each idea.

Pipelined Indirect Indexing: The first algorithm is based on indirect indexing. According to the sparse distribution implied in the current routing table, the number of entries can be reduced by just adding few fields. Furthermore, a new compression scheme is presented to solve the structural hazard, thus can be implemented in hardware pipelining. The experimental results demonstrate that the table size is small enough to fit into SRAM. The scheme cannot support incremental update and future IPv6 lookup, however, its implementation is relatively simple.

Binary Search on Routing Intervals: The major complexity of IP address lookup ties to the BMP problem. It is observed that the route aggregation implied in the Internet causes a special property called "routing locality". By adopting this property, the routing prefix is transformed to "**Routing Interval**".

With the routing interval, the complexity of BMP algorithms can be removed. The routing interval can be applied to most existing IP lookup algorithms. With binary search, the address lookup speed can outperform the existing scheme.

LS trie: Trie-based algorithm is the most straight-forward solution for address lookup. By introducing the concept "Level Smart-Compression", the advantages of LC-tries and binary trie are combined. With the LS trie, the size of the forwarding table is significantly reduced and the incremental updates is supported. It also provides the flexibility by separating the internal nodes and leaves to conform the memory hierarchy.

Tuple Reduction: Last, we extend the algorithm from IP address lookup to the policy-based routing. To solve the two-dimensional filters lookup, the tuple reduction algorithm uses the pre-computation to improve the lookup speed and storage over previous scheme simultaneously. Several heuristics are also provided to improve the lookup performance.

Recently, the research focus has been moved from IP address lookup and packet classification to deeper packet inspection by filtering predefined strings in the packet payload. The applications of deep packet inspection include virus detection, intrusion detection and data secrecy. Since these problems, including IP address lookup, packet classification and deep packet inspection, can be transformed into the problem of computational geometry, we believe that the advance in IP address lookup and packet classification could benefit to the solution of deep packet inspection.

References

[1] A Tam. How to Survive as an ISP. In: *Networld Interop*; 1997. .

[2] S Keshav and R Sharma. Issues and Trends in Router Design. *IEEE Communication Magazine* 1998;36(5):144–151.

[3] C Partridge, *et al*. A 50-Gb/s IP Router. *IEEE/ACM Trans On Networking* 1998;6(3):237–248.

[4] IETF MPLS Charter. Multiprotocol Label Switching See *http://www.ietf.org/html-charters/mpls-charter.html*, 1997;.

[5] AM Odlyzko. Data Networks are mostly empty and for good reason. In: *IT Professional*; 1999. .

[6] Y Rekhter, T Li. *An Architecture for IP Address Allocation with CIDR*. RFC 1993;(1518).

[7] C Partridge. Locality and Route Caches. In: *NSF Workshop on Internet Statistics Measurement and Analysis*; 1996. .

[8] NLANR Project. National Laboratory for Applied Network Research. See *http://www.nlanr.net*;.

[9] M Degermark, A Brodnik, SCarlsson, and S Pink. Small Forwarding Tables for Fast Routing Lookups. In: *ACM SIGCOMM*; 1997. p. 3–14.

[10] P Gupta, S Lin, and N McKeown. Routing Lookups in Hardware at Memory Access Speeds. In: *IEEE INFOCOM*; 1999. p. 1240–1247.

[11] N Huang, S Zhao and J Pan. A Novel IP-Routing Lookup Scheme and Hardware Architecture for Multigigabit Switching Routers. *IEEE JSAC* 1999;17(6):1093–1104.

[12] V Srinivasan and G Varghese. Fast IP lookups using controlled prefix expansion. *ACM Trans On Computers* 1999;17:1–40.

[13] S Nilsson and G Karlsson. IP-Address Lookup Using *LC − Tries*. *IEEE JSAC* 1999;17(6):1083–1092.

[14] M Waldvogel, G Varghese, J Turner and B Plattner. Scalable High Speed IP Routing Lookups. In: *ACM SIGCOMM*; 1997. p. 25–36.

[15] B Lampson, V Srinivasan and G Varghese. IP Lookups Using Multiway and Multicolumn Search. *IEEE/ACM Trans On Networking* 1999; 7(4):323–334.

[16] Pi-Chung Wang, Chia-Tai Chan and Yaw-Chung Chen. A Fast IP Lookup Scheme for High-Speed Networks. *IEEE Communications Letters* 2001; 5(3):125–127.

[17] Merit Networks Inc. Internet Performance Measurement and Analysis (IPMA) Statistics and Daily Reports. *IMPA Project* See *http* : //*www.merit.edu*/*ipma*/*routing_table*/;.

[18] Pi-Chung Wang, Chia-Tai Chan and Yaw-Chung Chen. Performance Enhancement of IP forwarding by using Routing Interval. *Journal of Communications and Networks* 2001;3(4):374–382.

[19] Pi-Chung Wang, Chia-Tai Chan and Yaw-Chung Chen. Performance Enhancement of IP Forwarding by Reducing Routing Table Construction Time. *IEEE Communications Letters* 2001;5(5):230–232.

[20] RP Draves, C King, S Venkatachary and BD Zill. Constructing Optimal IP Routing Tables. In: *IEEE INFOCOM*; 1999. p. 88–97.

[21] R Stevens and G Wright. *TCP/IP Illustrated-Vol. 2: THe implementation*. ; 1995. .

[22] K Sklower. *A tree-based table for Berkeley unix*.; 1991. .

[23] A Andersson and S Nilsson. Improved behavior of tries by adaptive branching. *Information Processing Letters* 1993;46(6):295–300.

[24] Pi-Chung Wang, Chia-Tai Chan, Shuo-Cheng Hu, Chung-Liang Lee and Wei-Chun Tseng. High-Speed Packet Classification for Differentiated Services in NGNs. *IEEE Transactions on Multimedia* 2004;6(6):925–935.

[25] Pankaj Gupta and Nick McKeown. Packet Classification on Multiple Fields. In: *ACM SIGCOMM*; 1999. p. 147–160.

[26] V Srinivasan, G Varghese and S Suri. Packet Classification using Tuple Space Search. In: *ACM SIGCOMM*; 1999. p. 135–146.

[27] Daxiao Yu, Brandon C Smith, and Belle Wei. Forwarding Engine For Fast Routing Lookups and Updates. In: *IEEE Globecom*; 1999. p. 1556–1564.

[28] Pankaj Gupta and Nick McKeown. Algorithms For Packet Classification. *IEEE Network Magazine* 2001;15(2):24–32.

[29] V Srinivasan, G Varghese, S Suri and M Waldvogel. Fast Scalable Level Four Switching. In: *ACM SIGCOMM*; 1998. p. 191–202.

[30] TV Lakshman and D Stidialis. High Speed Policy-based Packet Forwarding Using Efficient Multi-dimensional Range Matching. In: *ACM SIGCOMM*; 1998. p. 203–214.

[31] Anja Feldmann and S Muthukrishnan. Tradeoffs for Packet Classification. In: *IEEE INFOCOM*; 2000. p. 1193–1202.

[32] M Buddhikot, S Suri and M Waldvogel. Space Decomposition Techniques for Fast Layer-4 Switching. In: *IFIP Sixth International Workshop on High Speed Networks*; 2000. .

[33] Thomas Woo. A Modular Approach to Packet Classification: Algorithms and Results. In: *IEEE INFOCOM*; 2000. p. 1213–1222.

[34] Pankaj Gupta and Nick McKeown. Packet Classification using Hierarchical Intelligent Cuttings. In: *Hot Interconnects VII*; 1999.

Index

A

access, 1, 4, 8, 9, 10, 11, 16, 18, 19, 34, 35, 48, 49, 50, 63, 65, 66, 67
ACM, 79, 80, 81
ad hoc, 52
administrative, 53
administrators, 54
age, 34
aggregation, 4, 77
algorithm, 11, 12, 13, 14, 15, 16, 17, 24, 27, 28, 29, 32, 38, 41, 44, 46, 47, 48, 52, 53, 55, 56, 57, 60, 61, 62, 68, 69, 70, 71, 72, 74, 75, 77, 78
assumptions, 52

B

bandwidth, vii, 1, 2, 3, 11, 53
behavior, 7, 80
bottleneck, vii, 67
branching, 8, 41, 80
buses, 67

C

cache, 7, 8, 27, 32, 34, 35, 36, 38, 49, 51, 56, 64, 65, 66, 67, 71, 72
CAM, 5
capacity, 3
category d, 58
channels, 1
children, 8, 41, 44
classes, 3, 53, 54, 55
classification, 53, 54, 55, 56, 64, 67, 72, 74, 78
clouds, 22
collisions, 72
complexity, vii, 8, 9, 11, 21, 23, 26, 32, 34, 38, 46, 55, 56, 57, 61, 64, 77, 78
components, vii
computation, 13, 21, 64
construction, 12, 14, 15, 27, 28, 35, 38, 44, 47
consumption, vii, 56
costs, 32, 64
coupling, 17
CPU, 21, 32, 34, 35, 36, 37, 38, 49

D

data structure, viii, 9, 23, 27, 28, 34, 36, 37, 39, 41, 43, 44, 49, 51, 52, 77
database, 8, 54, 55, 58, 62, 68, 70, 71, 75
decisions, 2
decoupling, 39, 51, 52
definition, 1
delivery, 1
demand, 1
density, 22, 62, 63
Department of Defense, 1
designers, 7
detection, 78
differentiation, 2, 53, 54
disaster, 1
distribution, 4, 11, 14, 15, 22, 26, 32, 52, 58, 68, 75, 77
duplication, 41, 58

Index

E

e-commerce, 1
end-to-end, vii
engines, vii
enterprise, 1, 22
environment, 35
expansions, 61, 75

F

fiber, vii, 3
filters, 54, 55, 56, 57, 58, 59, 60, 61, 62, 63, 64, 65, 68, 69, 70, 78
flexibility, 2, 53, 56, 78
flow, 54, 56

G

generalization, 3, 56
generation, 21, 69
grids, 63
groups, 57
growth, 7

H

handling, 38, 74
height, 28, 35
heme, 73, 75
heuristic, 52, 55, 60, 61, 63
high-level, 16
high-speed, vii, viii, 18, 70, 72, 77
host, 3, 5

I

IETF, 3, 79
implementation, viii, 9, 16, 21, 23, 29, 35, 36, 40, 55, 66, 67, 70, 73, 77, 80
indexing, 14, 77
infrastructure, 53
insertion, 29, 36, 41, 42, 48, 49, 51
inspection, 78
Internet, vii, 1, 2, 3, 5, 7, 22, 54, 55, 58, 75, 77, 79, 80
Internet Protocol, vii, 1, 2, 3, 4, 5, 7, 8, 9, 10, 11, 13, 15, 16, 17, 18, 19, 21, 22, 23, 25, 26, 27, 28, 29, 32, 33, 35, 37, 38, 39, 40, 41, 42, 43, 45, 47, 48, 49, 51, 53, 54, 55, 56, 57, 62, 77, 78, 79, 80
interval, 23, 24, 26, 27, 28, 29, 30, 32, 33,
IP address, 2, 3, 4, 5, 8, 9, 10, 11, 17, 21, 22, 23, 32, 35, 38, 53, 55, 56, 62, 77, 78
IP networks, 53
IPv4, vii, 3, 8, 10, 39, 49
IPv6, vii, 5, 8, 32, 39, 77
ISPs, 1, 54
iteration, 62

K

King, 80

L

limitation, vii, 24
linear, 8, 28, 55, 56, 57, 69, 70
links, vii, 1, 3, 5, 71
location, 21, 65, 66

M

management, 69
mapping, 2, 32, 44, 53, 60
measures, 34
memory, vii, 8, 9, 10, 11, 14, 16, 17, 18, 19, 28, 32, 34, 35, 37, 38, 39, 40, 42, 49, 50, 52, 56, 57, 65, 66, 67, 68, 69, 70, 77, 78
misleading, 26
multidimensional, 57
multimedia, vii, 1, 54, 74
multimedia services, 54, 74

N

natural, 1
network, vii, 1, 3, 4, 5, 7, 22, 26, 42, 48, 54, 67
nodes, 23, 28, 39, 40, 41, 42, 43, 44, 46, 47, 48, 49, 51, 52, 78

O

observations, 22
one-to-one mapping, 44
optimization, 55, 62, 63, 69, 70, 71, 72, 75
optimization method, 55
organization, 7, 14

P

packet forwarding, vii, 1, 2, 3, 53
packets, vii, 1, 2, 3, 7, 21, 38, 48, 53, 54, 55, 56, 70
parallel implementation, 67, 73
parallelism, 56, 77
parameter, 45
partition, 29
performance, vii, viii, 3, 11, 21, 23, 27, 28, 32, 34, 35, 36, 37, 39, 40, 41, 43, 49, 50, 51, 54, 55, 56, 57, 58, 66, 68, 69, 70, 71, 72, 74, 78
permit, 74
personal, 1
personal computers, 1
pipelining, 16, 19, 77
PNA, 54
point-to-point, 5
poor, 28, 54, 68
poor performance, 28
ports, 14, 22
power, vii, 3, 4, 27, 56
preferential treatment, 2, 53
probability, 72
probe, 57, 58, 66
programming, 8, 27, 41, 55, 58, 60, 61, 64, 75
promote, 58
property, 22, 77
protocol, vii, 2, 53, 54
protocols, vii, 1, 3, 42
pseudo, 44
public, 1, 5

Q

QoS, 2, 53, 54

R

random, 69
range, 13, 15, 54
reconstruction, 32, 35
reduction, 55, 59, 60, 63, 68, 69, 70, 72, 73, 75, 78
regular, 75
research, vii, 78
resource allocation, 54
resources, 53
returns, 57, 58, 65
routing, vii, 2, 3, 4, 5, 7, 8, 9, 11, 12, 13, 15, 17, 18, 21, 22, 23, 24, 25, 26, 27, 28, 29, 30, 32, 34, 35, 36, 37, 38, 39, 40, 42, 43, 44, 45, 46, 47, 48, 50, 52, 53, 58, 62, 68, 77, 78, 80

S

sample, 4, 11, 12, 27, 43
scalability, 32, 39, 56, 70, 77
scalable, 55, 77
search, vii, 5, 8, 21, 24, 26, 27, 28, 32, 33, 34, 35, 36, 43, 48, 51, 55, 56, 57, 58, 59, 62, 63, 64, 65, 66, 69, 70, 71, 72, 73, 74, 75, 78
searches, 8
searching, 27, 57, 58, 68
secrets, 68
security, 2, 53
selecting, 64
services, 53, 54, 64, 67, 74
simulation, 52
simulations, 7
sites, 2, 53
software, viii, 7, 8, 9, 21, 56, 77
solutions, 3, 7, 8, 55, 56
sorting, 22, 32
spatial, 55, 56
speed, vii, 3, 7, 17, 34, 37, 38, 41, 43, 49, 50, 51, 52, 55, 63, 65, 68, 69, 70, 71, 72, 75, 77, 78
storage, vii, 23, 28, 35, 42, 43, 50, 51, 55, 58, 59, 60, 62, 63, 64, 71, 72, 74, 75, 77, 78
streams, 40
structuring, 52
subnetworks, 1
switching, 2, 3

systems, 1

T

targets, 17
TCP, 2, 53, 80
TCP/IP, 80
technology, vii, 3, 9
temporal, 56
threshold, 63
time, 8, 11, 14, 28, 32, 35, 36, 37, 38, 40, 41, 42, 49, 50, 57, 61, 65, 66, 67, 69, 70, 71, 72, 73, 75
topology, 22, 26, 48
tracking, 57
trade, 12, 27
trade-off, 17, 70
trading, 62
traffic, 1, 2, 3, 35, 53
traffic flow, 35, 53
transfer, 1
transformation, 24, 28, 29, 32, 35, 36, 38
transmission, vii, 1, 3
tree-based, 80

two-dimensional, 55, 57, 59, 78

U

updating, 11, 48

V

values, 16, 29, 46, 49
variable, vii, 5, 11, 44, 49, 54
variation, 12
vector, 15
virus, 78
visible, 4
VoIP, 54

W

war, 1
World Wide Web, 1